Advancing Recommender Systems with Graph Convolutional Networks

Fan Liu • Liqiang Nie

Advancing Recommender Systems with Graph Convolutional Networks

Springer

Fan Liu
School of Computing
National University of Singapore
Singapore, Singapore

Liqiang Nie
Harbin Institute of Technology
Shenzhen, China

ISBN 978-3-031-85092-9 ISBN 978-3-031-85093-6 (eBook)
https://doi.org/10.1007/978-3-031-85093-6

This Springer imprint is published by the registered company Springer Nature Switzerland AG
The registered company address is: Gewerbestrasse 11, 6330 Cham, Switzerland

Preface

Recommender systems have become an indispensable component of modern online platforms, shaping user experiences by offering personalized suggestions that enhance satisfaction, increase engagement, and drive business success. From streaming services to e-commerce platforms, these systems leverage vast amounts of data to anticipate user needs and provide tailored content. Among the many advancements in this domain, Graph Convolutional Networks (GCNs) have emerged as a transformative tool. By effectively modeling graph structures, GCNs excel in capturing high-order relationships between users and items, enabling the creation of enriched and expressive representations. However, despite their promise, both GCN-based methods and recommender systems at large face critical challenges that constrain their scalability, effectiveness, and adoption.

This book systematically examines and addresses these challenges, categorized into two overarching themes. The first focuses on problems specific to GCN-based recommendation models, including:

1. **Over-smoothing Problem**: A pervasive problem where user and item embeddings converge to indistinguishable representations as GCN layers deepen, diminishing personalization and recommendation quality.
2. **Noisy Neighboring Nodes Issue**: GCN models often aggregate information indiscriminately from all high-order neighbors, potentially amplifying irrelevant or unreliable signals, thereby compromising the quality of recommendations.
3. **Interpretability limitations**: The intricate aggregation processes in GCNs obscure the reasoning behind recommendations, reducing transparency and trust in critical applications.

The second category encompasses broader challenges in recommendation systems that GCN-based methods are particularly well-suited to address:

1. **Attribute missing problem**: Many recommendation datasets lack complete side information, such as item attributes, which limits the ability to accurately model user preferences.

2. **Feature misalignment**: Multimodal recommendation systems often struggle with aligning extracted features to user preferences, leading to suboptimal performance in predicting interactions.

Through rigorous exploration of these challenges, this book presents innovative GCN-based solutions to push the boundaries of recommender system design. Techniques such as interest-aware message-passing strategy, cluster-based collaborative filtering, semantic aspects extraction, attribute-aware attention mechanisms, and light graph transformer are introduced to address the concerns of improving GCN models and solving broader issues in recommendation tasks. These approaches demonstrate how GCNs can be tailored to mitigate the over-smoothing problem and noisy neighboring nodes issue, provide interpretable recommendations, handle missing data, and align multimodal features in recommender systems, thereby empowering recommender systems to achieve superior performance.

Beyond advancing GCN-based recommendation methods, this book illustrates how GCN-based approaches can transform the broader landscape of recommendation systems, equipping them to handle the complex relationships and data imperfections inherent in real-world applications. Each chapter combines theoretical insights with practical implementations and experimental validation, offering a comprehensive resource for researchers, practitioners, and students alike. By systematically addressing these challenges, this book not only deepens our understanding of GCN-based recommendation systems but also highlights their potential to overcome key obstacles in modern recommender systems. We hope it serves as an inspiration for future innovation, guiding readers to design systems that are not only more effective but also more transparent and reliable.

Singapore, *Fan Liu*
Liqiang Nie

Acknowledgements

This book would not have been completed, or at least not be what it looks like now, without the invaluable support of many colleagues, particularly those from the N-CRiPT Lab at National University of Singapore and the iLearn Center at Shandong University. We are deeply grateful for their contributions to this time-intensive yet highly meaningful project. Their dedication, thoughtful insights, and significant input in the discussions have greatly enriched the writing process and shaped the final outcome of this work.

Our first and foremost thanks undoubtedly go to colleagues who are also the contributors of some chapters of this book: Prof. Mohan Kankanhalli at National University of Singapore, Prof. Tat-Seng Chua at National University of Singapore, Prof. Zhiyong Cheng at Hefei University of Technology, Prof. Weili Guan at Harbin Institute of Technology (Shenzhen), Prof. Yinwei Wei at Shandong University, Dr. Zhenyang Li at Hong Kong University of Science and Technology and Mr. Huilin Chen at Hefei University Of Technology. Thanks for their active participation in the technical discussion of this book and their constructive feedback and comments that have been significantly helpful in shaping this book.

Second, we are very grateful to the anonymous reviewers for taking the time to read the book despite their busy schedules. Their insightful and constructive feedback significantly enhanced the clarity and readability of this book.

Third, we sincerely extend our thanks to the Executive Editor, Mr. Ralf Gerstner, and the Product Editor, Ms. Ramya Prakash, whose support greatly contributed to making the book publishing process smooth and enjoyable.

Last, but certainly not least, our thanks are reserved to our beloved families for their selfless consideration, and unconditional support. We dedicate this book to them, with love.

Fan Liu and Liqiang Nie
January 2025

Contents

Authors' Biographies

Dr. Fan Liu is a Research Fellow with the School of Computing, National University of Singapore (NUS). He received his Ph.D. from Shandong University, China. His research interests lie primarily in multimedia information retrieval and recommendation. His work has been published in a set of top forums, including ACM SIGIR, MM, WWW, T-KDE, T-oIS, T-MM, and T-CSVT. He has served as an Area Chair for ACM MM, a Senior Program Committee member for CIKM, and a Guest Editor for T-oMM. Meanwhile, he has served as a Technical Program Committee member for several top conferences, like ACM SIGIR, WWW, and ICML, and a reviewer for journals including T-KDE, T-oIS, T-MM, and T-CSVT.

Dr. Liqiang Nie, fellow of IAPR, is the director with the College of Information Science and Technology, Harbin Institute of Technology (Shenzhen campus). He received his Ph.D. degree from National University of Singapore. His research interests are multimedia content analysis and information retrieval. Dr. Nie has co-/authored more than 150 papers and 5 books, with 30,000 plus Google citations. He is an AE of IEEE T-KDE, IEEE T-MM, IEEE T-CSVT, ACM T-oMM, and Information Science. Meanwhile, he serves as the chair of ICMR 2025, ICME 2025, and ACM MM 2027. He is a member of ICME steering committee. He has received many awards over the past four years, like SIGMM rising star in 2020, MIT TR35 China 2020, SIGIR best student paper in 2021, IEEE AI's 10 to Watch in 2022, and ACM MM Best paper award in 2022.

Acronyms

BPR Bayesian Personalized Ranking
CE Cross Entropy
CF Collaborative Filtering
CNN Convolutional Neural Network
DNN Deep Neural Network
GCN Graph Convolutional Network
LLM Large Language Model
MF Matrix Factorization
MLP Multi-Layer Perceptron
NDCG Normalized Discounted Cumulative Gain
RS Recommender System

Chapter 1
Introduction

1.1 Background

Recommendation systems have become a cornerstone of various online platforms, including e-commerce websites, news portals, and social media sites. These systems are pivotal in enhancing user experiences by helping individuals navigate the overwhelming abundance of information to discover content that aligns with their interests. Simultaneously, they play a crucial role in boosting revenue for service providers by fostering user engagement and improving conversion rates. Platforms like Amazon [1], eBay [2], Mercari [3], Netflix [4], YouTube [5], and TikTok [6] effectively leverage recommendation systems to connect users with relevant products, videos, or entertainment. Similarly, social platforms like Facebook [7] and Instagram [8] employ personalized recommendations to enhance user interaction and content discovery, while news portals like Flipboard [9] and SmartNews [10] curate articles tailored to user preferences. With these platforms hosting millions to billions of items—ranging from products and videos to music, news, and social connections—recommendation systems have become indispensable for personalizing user experiences and optimizing platform performance. Their effectiveness directly influences platform revenue, making them a key area of focus for both academia and industry.

Within this field, Collaborative Filtering (CF)-based models have achieved significant advancements in learning the user and item representations by effectively

[1] https://www.amazon.com

[2] https://www.ebay.com

[3] https://www.mercari.com

[4] https://www.netflix.com

[5] https://www.youtube.com

[6] https://www.tiktok.com

[7] https://www.facebook.com

[8] https://www.instagram.com

[9] https://www.flipboard.com

[10] https://www.smartnews.com

F. Liu, L. Nie, *Advancing Recommender Systems with Graph Convolutional Networks*,
https://doi.org/10.1007/978-3-031-85093-6_1

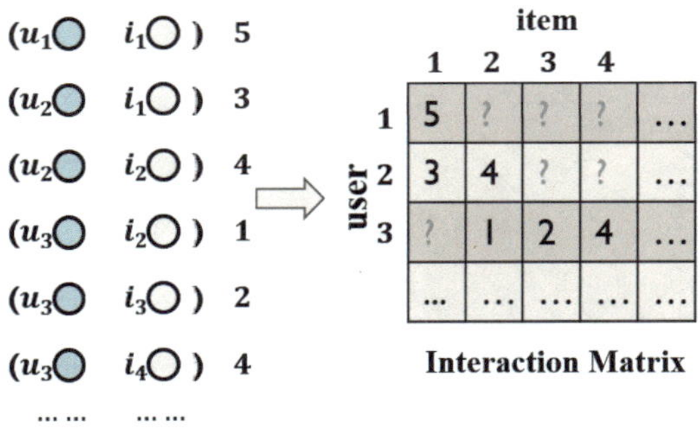

Fig. 1.1: An example of a user-item interaction matrix, which MF-based methods utilize to learn user and item representations by reconstructing its structure.

exploiting the historical user-item interactions [4, 56, 148, 154]. The core idea of CF lies in the assumption that users with similar interaction histories are likely to share similar preferences, and this assumption has been the basis for various modeling techniques. A classic example is Matrix Factorization (MF), which first models the user-item interaction matrix (as illustrated in Fig. 1.1) as a product of two lower-dimensional latent factor matrices, representing users and items in a shared embedding space. Then, it predicts the user preference with the inner product [4]. This linear modeling approach, while effective, struggles to capture complex user-item interaction behaviors. To address these limitations, neural network-based CF models have been introduced, significantly advancing the field. One notable example is NeuMF (Neural Matrix Factorization) [56], which integrates the strengths of both generalized linear models and nonlinear neural networks. NeuMF uses a multilayer perceptron (MLP) to model non-linear and complex interaction patterns, capturing intricate user behaviors that traditional MF cannot.

Despite significant progress, MF-based methods model only direct interactions and cannot capture high-order connectivity, restricting their ability to learn rich user-item relationships. More recently, Graph Convolution Networks (GCNs)-based models [5, 136, 135, 54, 88, 138, 87, 37, 141] offer a cutting-edge approach in recommendation systems by leveraging high-order connectivities based on the topological structure of the interaction data. As shown in Fig. 1.2, these models obtain the final node representations ($a.k.a$ user and item embeddings) by iteratively aggregating information from neighboring nodes. Many methods have demonstrated that exploring additional information based on graph topology structure can efficiently enhance user and item representation, ultimately improving recommendation performance. For example, NGCF [136] incorporates high-order connectivity of user-item interactions in the user-item bipartite graph for representation learning in recommendation. With careful experimental studies, He et al. [54] simplified NGCF and achieved better performance by removing the feature transformation and the non-linear activation module. This is because these two components do not contribute positively to the final performance and may even have a negative effect due to their impact on training

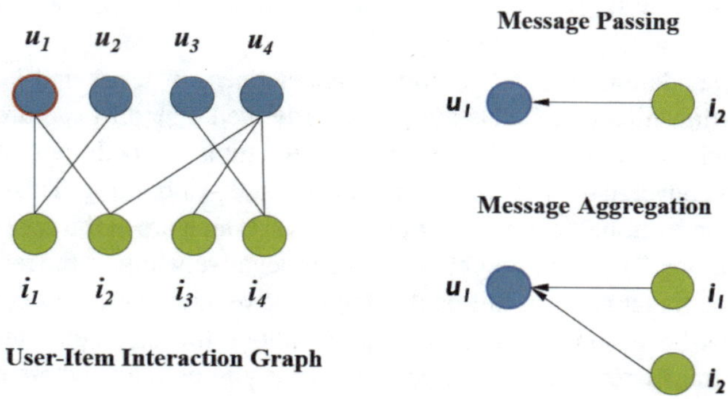

Fig. 1.2: An example of user-item interaction graph. The message passing and message aggregation are two fundamental operations of the graph convolutional network. The target node u_1 aggregates information from its neighboring nodes to derive its final representation.

difficulty. Furthermore, GCN-based methods hold great potential in addressing key issues in recommendation systems by capturing high-order connectivity to model complex user-item relationships and addressing data sparsity through information propagation. Their ability to integrate multi-modal and side information also makes them versatile tools for improving recommendation performance and adapting to various scenarios.

In summary, the application of GCNs in recommender systems has been highly successful, driving significant advancements by leveraging high-order connectivity and effectively modeling complex user-item relationships. These methods have proven adept at addressing key challenges, such as sparse interaction data and the need for richer representations, establishing themselves as a cornerstone of modern recommendation research. However, GCNs are not without limitations, which must be addressed to fully unlock their potential. Notably, they hold significant promise for further innovation, particularly in integrating additional information. This ongoing evolution positions GCN-based models as a powerful catalyst for the continued advancement and refinement of recommender systems.

1.2 Motivations

Graph Convolutional Networks (GCNs) have emerged as a transformative tool in recommender systems by leveraging graph structures to model intricate user-item interactions. However, their potential is often hindered by challenges specific to GCNs in recommendation tasks, such as the over-smoothing problem, the noisy neighboring nodes issue, and interpretability limitations, as well as broader challenges in recommendation systems, including missing attributes and feature misalignment, which GCNs could help mitigate.

In recommendation systems, the interconnectedness of users and items creates complex relationships that GCNs are particularly well-suited to capture. However, as GCNs stack layers to model high-order interactions, embeddings of dissimilar nodes tend to converge—a phenomenon known as over-smoothing. This undermines personalization by homogenizing user preferences, which are at the core of effective recommendations. The noisy neighboring nodes negatively impact the representation learning of the target node. Additionally, the iterative aggregation process in GCNs often obscures the reasoning behind recommendations, limiting their interpretability. This lack of transparency reduces trust, particularly in domains where explainability is critical. Beyond these GCN-specific challenges, recommender systems face broader issues that GCNs can address. Attributes play a vital role in enhancing user and item embeddings, particularly in sparse interaction scenarios. Yet, real-world datasets frequently suffer from missing or incomplete attribute information, which hinders model effectiveness. Furthermore, multimodal recommendation systems often grapple with feature misalignment, where extracted features fail to align with the factors driving user preferences. GCNs, with their ability to model relationships and incorporate contextual information, offer promising solutions to these challenges.

This book systematically addresses these challenges by introducing innovative methods tailored to GCN-based recommendation systems. These methods aim to optimize user and item representation learning, improve interpretability, and integrate multimodal data effectively, addressing both the intrinsic limitations of GCNs and the broader challenges in recommendation tasks. By combining theoretical insights with practical implementations, this book seeks to inspire future research and equip practitioners with robust tools to create more accurate, interpretable, and personalized recommender systems, ultimately pushing the boundaries of what GCNs can achieve in this evolving field.

1.3 Challenges

This section categorizes key challenges in recommendation systems with graph convolutional networks into two classes. The first class focuses on challenges specific to GCN-based recommendation models, including the over-smoothing problem, where embeddings for users and items become overly similar with deeper layers; the issue of the noisy neighboring nodes, which aggregate unreliable information negatively impacts the representation of target node; and the interpretability limitations, which obscure the reasoning behind recommendations and undermine trust. The second class highlights broader challenges in recommendation systems that could be mitigated by GCN-based approaches, such as the attribute-missing problem, where incomplete contextual information limits model effectiveness, and the challenge of feature misalignment, where irrelevant or misaligned features hinder accurate user preference modeling. These challenges collectively underscore the need for innovations that enhance both the functionality and reliability of recommendation systems.

1.3.1 Over-smooting Problem

Graph Convolutional Networks (GCNs) have transformed recommendation systems by utilizing high-order connectivity and graph topology to model user-item interactions effectively. However, they are prone to the over-smoothing problem, where the embeddings of nodes in the graph—users and items—become overly similar as the number of layers in the GCN increases. This phenomenon poses a significant challenge in recommendation scenarios, as it undermines the core goal of personalization.

In the context of recommendation, over-smoothing results in the homogenization of user preferences, eroding the distinction between individuals with diverse or even contradictory interests. When GCNs stack multiple layers to capture high-order relationships, embeddings of dissimilar users—who may have distinct or unrelated preferences—begin to converge, diluting the unique aspects of their interactions. This homogenization leads to suboptimal recommendations, where the model fails to reflect users' personalized needs.

1.3.2 Noisy Neighboring Nodes

The noisy neighboring nodes problem in GCN-based recommendation systems arises when unreliable information from loosely connected or unrelated neighboring nodes negatively impacts the representation of target nodes. This occurs because GCN models typically aggregate information indiscriminately from all high-order neighbors without distinguishing their reliability or relevance. For instance, consider a user u_1 who prefers mystery novels. A high-order neighbor u_2, loosely connected to u_1, prefers gardening books. In a GCN-based system, u_2's preference for gardening books might influence the representation of u_1, diluting its relevance to mystery novels. This misrepresentation leads to inaccurate predictions and undermines the effectiveness of the recommendation system. Addressing the noisy neighboring nodes problem is essential to enhance the accuracy and reliability of GCN-based recommendation systems, enabling them to provide precise and personalized recommendations that align closely with user preferences.

1.3.3 Interpretability Limitations

Graph Convolutional Network (GCN)-based recommendation models have significantly enhanced recommendation systems by utilizing the graph structure of user-item interactions to model complex relationships. However, these models face a critical challenge: the lack of interpretability. Interpretability is crucial for understanding the factors driving user preferences and item recommendations, particularly in domains requiring transparency and trust, such as e-commerce, healthcare, and personalized content delivery.

The core of GCN-based models lies in their iterative aggregation of information from neighboring nodes in the user-item interaction graph. This aggregation process enables the capture of intricate, high-order relationships between users and items, thereby improving recommendation accuracy. However, it also inherently obscures the reasoning behind recommendations. The multi-layer propagation and non-linear transformations in GCNs make it difficult to disentangle specific contributions of individual user-item interactions or features to the final recommendations. Consequently, neither users nor service providers can gain a clear understanding of the "why" behind the suggested items, which reduces trust in the system and hampers its adoption in scenarios where interpretability is non-negotiable.

1.3.4 Attribute Missing Problem

The attribute missing problem presents a major challenge in building recommendation models. Attributes, such as the genre or cast of movies and the category or brand of products, are essential for enriching user-item interaction data and improving recommendation accuracy. These attributes provide contextual information that helps mitigate the sparsity inherent in interaction data, a common limitation in collaborative filtering-based methods. However, in practical scenarios, attribute information is often incomplete, with many items lacking values for certain attributes.

This missing attribute issue disrupts the effectiveness of recommendation models. Attributes that are unavailable or incomplete result in gaps in the data that the model relies on to understand and represent items and users. As a result, the model's ability to learn accurate embeddings and make precise recommendations is hindered. Moreover, the absence of attributes can lead to biased or less meaningful representations, exacerbating the challenges posed by data sparsity. This problem is pervasive in real-world systems, where comprehensive and consistent attribute information is rarely guaranteed.

1.3.5 Feature Misalignment

Effective multimodal feature extraction remains a significant challenge in recommendation systems, as it directly affects the accuracy of user preference modeling and interaction prediction. Most existing approaches follow a standard pipeline consisting of three stages: multimodal feature extraction, user preference modeling, and user-item similarity measurement. While advancements such as deep neural networks and graph convolutional networks have brought improvements, they often overlook the alignment between extracted features and the factors that genuinely drive user preferences.

A key issue lies in the reliance on pre-trained feature extractors, which are primarily designed to capture discriminative signals for item categorization rather than preference-related attributes. For example, features like shoelaces may effectively differentiate sneakers from dress shoes but are less relevant than attributes like brand or style for predicting user preferences. Consequently, these extractors often produce features that include irrelevant or misaligned information, reducing the effectiveness of subsequent user preference modeling and impairing recommendation performance.

1.4 Our solutions

In this book, to tackle the aforementioned research challenges, we present some state-of-the-art graph convolutional networks tailored to recommender systems. Our approaches systematically enhance the capability of recommendation systems. To validate our proposed methods, we conducted comprehensive experiments on publicly available datasets, demonstrating significant improvements and robust performance across various scenarios. The results indicate that our techniques are well-suited to tackle the inherent complexities of graph convolutional networks and the problem in recommendation, thereby advancing the field of recommender systems.

1.4.1 Interest-aware Message-Passing Graph Convolutional Network

Like other GCN models, the GCN-based recommendation models also suffer from the smoothing problem. The proposed LightGCN and LR-GCN alleviate this problem to some extent, however, we argue that they overlook an important factor for the over-smoothing problem in recommendation, that is, high-order neighboring users with no common interests of a user can be also involved in the user's embedding learning in the graph convolution operation. As a result, the multi-layer graph convolution will make users with dissimilar interests have similar embeddings. To mitigate the over-smoothing problem, we propose a novel Interest-aware Message-Passing GCN (IMP-GCN) recommendation model, which performs high-order graph convolution inside subgraphs. The subgraph consists of users with similar interests and their interacted items. To form the subgraphs, we design an unsupervised subgraph generation module, which can effectively identify users with common interests by exploiting both user features and graph structure. To this end, our model can avoid propagating negative information from high-order neighbors into embedding learning.

1.4.2 Cluster-based Graph Collaborative Filtering Model

Noisy information from unreliable neighboring nodes (*e.g.*, users with dissimilar interests) adversely affects the representation learning of the target node. We argue that most existing GCN-based methods fail to account for users' multiple interests when performing high-order graph convolution. To address this limitation and enhance representation learning by extracting more valuable information from high-order neighbors while mitigating noise, we propose a novel GCN-based recommendation model, Cluster-based Graph Collaborative Filtering (ClusterGCF). This model performs high-order graph convolution on cluster-specific graphs, which are constructed by capturing users' multiple interests and identifying shared interests among them. Specifically, we introduce an unsupervised and optimizable soft node clustering method to categorize user and item nodes into multiple clusters. Using the soft clustering results and the topology of the user-item interaction graph, we assign nodes to clusters with probabilistic weights, enabling the construction of cluster-specific graphs. This approach effectively leverages the diversity of user interests while reducing the influence of noisy neighbors, leading to improved recommendation performance.

1.4.3 Semantic Aspect-based Graph Convolutional Network

To address the challenge of improving the interpretability of GCN-based recommendation systems, we introduce an approach that integrates the LLMs with traditional recommendation methods. Previous studies improve recommendation accuracy and interpretability by modeling user preferences across various aspects and intents. However, the aspects and intents are inferred directly from user reviews or behavior patterns, suffering from the data noise and the data sparsity problem. Furthermore, it is difficult to understand the reasons behind recommendations due to the challenges of interpreting implicit aspects and intents.

To overcome these limitations, we harness the sentiment analysis capabilities of LLMs to enhance the accuracy and interpretability of the conventional recommendation methods. Specifically, inspired by the deep semantic understanding offered by LLMs, we introduce a chain-based prompting strategy to uncover semantic aspect-aware interactions, which provide clearer insights into user behaviors at a fine-grained semantic level. To incorporate the rich interactions of various aspects, we propose a simple yet effective Semantic Aspect-based Graph Convolution Network. By performing graph convolutions on multiple semantic aspect graphs, SAGCN efficiently combines embeddings across multiple semantic aspects for final user and item representations.

1.4.4 Attribute-aware Attentive Graph Convolutional Network

To address the attribute-missing problem, we present an attribute-aware attentive graph convolution network (A^2-GCN). In particular, we first construct a graph, where users, items, and attributes are three types of nodes and their associations are edges. Thereafter, we leverage the graph convolution network to characterize the complicated interactions among <users, items, attributes>. Furthermore, to learn the node representation, we adopt the message-passing strategy to aggregate the messages passed from the other directly linked types of nodes (*e.g.*, a user or an attribute). Towards this end, we are capable of incorporating associate attributes to strengthen the user and item representation learning, and thus naturally solve the attribute missing problem. Given that for different users, the attributes of an item have different influence on their preference to this item, we design a novel attention mechanism to filter the message passed from an item to a target user by considering the attribute information.

1.4.5 Light Graph Transformer Model

To address the feature misalignment problem, we leverage the Transformer model to capture informative features by establishing correlations between items historically interacted with by the same user. Recognizing the challenges of effectiveness and efficiency associated with Transformer models, we propose a novel Transformer-based recommendation model, termed the Light Graph Transformer (LightGT). This model employs a Modal-Specific Embedding to enable accurate similarity measurement by tailoring embeddings to specific modalities, a Layer-Wise Position Encoder to enhance the modeling of sequential relationships across layers for improved representation of user-item interactions, and a Light Self-Attention Block to optimize the self-attention mechanism, reducing computational complexity while preserving scoring effectiveness.

1.5 Book Structure

In this book, we delve into the advancements of GCN-based methods in recommender systems, offering a detailed examination of cutting-edge models and techniques that harness the power of graph convolutional networks for personalized recommendations. This book is designed to benefit students, researchers, and practitioners who are interested in graph learning, recommender systems, and their intersections. It is worth emphasizing that the GCN-based methods discussed in this book are not confined to recommender systems but extend to other graph-related domains, such as knowledge graph reasoning, social network analysis, and molecular interaction prediction. By presenting a systematic exploration of innovative GCN-based approaches, this book equips readers with the knowledge to understand and apply these techniques to a wide range of challenges across various fields.

The remainder of this book comprises six chapters, each addressing a significant aspect of recommender systems research. Chapter 2 introduces the interest-aware message-passing graph convolutional network, which leverages high-order neighbors within the same subgraph. These user nodes exhibit more similar interests compared to those in other subgraphs, effectively mitigating the over-smoothing problem. In Chapter 3, we propose the cluster-based graph collaborative filtering model, which performs high-order graph convolution over cluster-specific graphs. This approach enhances the model's ability to filter out negative information while extracting valuable information from high-order neighboring nodes. Chapter 4 presents a semantic aspect-based graph convolutional network that conducts graph convolution operations on semantic aspect-aware interaction graphs. By incorporating semantic aspects, the model achieves improved interpretability. Chapter 5 explores the attribute-aware attentive graph convolutional network, which seamlessly integrates item attribute information into recommendations, naturally addressing the challenge of missing attributes. Chapter 6 focuses on a lightweight graph transformer model designed to extract valuable information from multimodal features for recommendation tasks. Finally, Chapter 7 offers a comprehensive summary of the book's key contributions and highlights promising directions for future research in recommender systems.

Chapter 2
Interest-aware Message-Passing Graph Convolutional Network

2.1 Introduction

As introduced in Chapter 1, Graph Convolutional Networks (GCNs) have been demonstrated to be an efficient way to distill additional information from graph structure, and thus improve user and item representation learning and alleviate the sparse problem. However, it is also well-recognized that GCNs suffer from the *over-smoothing* problem [136], because the graph convolution operation is actually a special kind of graph Laplacian smoothing [136], making node representations become indistinguishable after multi-layer graph convolution [153]. As a result, most current GCN-based models obtain their peak performance by stacking only a few layers (*e.g.*, 2 or 3 layers), and continuing to increase the depth will lead to sharp performance degradation. In the domain of recommendation, Chen et al. [16] have empirically demonstrated that the user/item embeddings become more similar when stacking more layers in NGCF due to the over-smoothing effect. In other words, the preferences of different users become homogeneous, resulting in performance degradation in recommendation. Based on the observations, they proposed the LR-GCN model [16], which removes the non-linearities in GCNs to simply the network structure and introduces a residual network structure to alleviate the over-smoothing problem, achieving substantial improvement over NGCF on recommendation accuracy.

It is worth mentioning that the LightGCN proposed by He et al. [54] has a similar formulation as LR-GCN. With careful experimental studies, He et al. pointed out that the feature transformation and nonlinear activation have no positive effect (or even negative effect due to the increase of training difficult) to the final performance. Therefore, they only keep the neighborhood aggregation in the LightGCN for collaborative filtering. Comparing to LR-GCN, LightGCN further removes the "self-loop" in the aggregation operation. Although LightGCN is not dedicatedly designed for tacking the over-smoothing problem, it has almost the same formulation as LR-GCN and thus can also alleviate the over-smoothing problem to some extent. In fact, both LR-GCN and LightGCN are consistent with the recent theories

F. Liu, L. Nie, *Advancing Recommender Systems with Graph Convolutional Networks*,
https://doi.org/10.1007/978-3-031-85093-6_2

in simplifying GCNs [145] and can obtain the best performance with a deeper structure (*e.g.*, 4 layers). Despite the two success GCN based models are designed for recommendation, we argue that they still design the model from the perspective of graph convolution, while have not well considered the over-smoothing problem in the domain of recommendation.

The GCN based recommendation model is built upon a user-item graph, in which the user and item are linked according to the historical user-item interactions. The user embedding is learned by iteratively aggregating messages passed from the neighboring (both user and item) nodes. Note that the passed messages are distilled from the embeddings of neighboring nodes. When stacking k layers, the information from the k-order neighbors, which are indirectly connected via items and users, are also involved in the embedding learning of a target node. An underlying assumption is that the collaborative signals from high-order neighbors are beneficial to the embedding learning. However, not all the information from high-order neighbors are positive in reality. In the user-item interaction graph, the high-order neighboring users could have no common or even contradictory interest with a target user. This is highly possible, especially when the graph is constructed based on implicit feedbacks (*e.g.*, click). In fact, the implicit feedback is more widely used over the explicit feedbacks in modern recommendation systems. The core idea behind collaborative filtering is that similar users like similar items. Therefore, the collaborative signals that we would like to exploit should be from similar users (*i.e.*, users with similar interests). However, existing GCN-based recommendation models have not distinguished the high-order neighbors, and just simply aggregate the messages from all those neighbors to update user embeddings. As a result, the embeddings of dissimilar users are also involved in the embedding learning of a target user, negatively affecting the performance. This is also a reason of the over-smoothing effect in the GCN-based recommendation models – making the embeddings of dissimilar users to be similar.

Motivated by the above considerations, in this chapter, we propose a novel Interest-aware Messaging-Passing GCN (IMP-GCN) recommendation model, which groups users and their interacted items into different subgraphs and operates high-order graph convolutions inside subgraphs. More specific, we adopt the simplified network structure of LightGCN, as its effectiveness has been well demonstrated in [54] and it can alleviate the over-smoothing problem to some extent. The first-order graph convolution is the same as that of LightGCN. For the high-order graph convolution, only the messages from nodes in the same *subgraph* are exploited to learn the node embeddings. The subgraph is generated by a proposed graph generation module, which integrates users features and graph structure to identify users with similar interests, and then constructs the subgraphs by retaining those users and their interacted items. To this end, our model can filter out the negative information propagation in the high-order graph convolution operations for the embedding learning, and thus can keep the uniqueness of users by stacking more graph convolution layers. Extensive experiments have been conducted on three large-scale real-world datasets to validate the effectiveness of our model. Results show that our model outperforms the state-of-the-art methods by a large margin and can obtain better performance with more layers (till 7 layers) [1]. This indicates that our model can benefit from higher-order neighbors by excluding negative nodes. Besides, with deep analysis of the results, we found that the negative information in the embedding propagation is the major reason for the performance degradation of existing GCN-based recommendation models in deep structure. We released the codes and involved parameter settings to facilitate others to repeat this method [2].

In summary, the main contributions of this work are as follows:

- We step into the over-smoothing problem in existing GCN-based recommendation models and point out an overlooked factor: exploiting high-order neighbors indiscriminately makes the embeddings of users with dissimilar interests to be similar.
- We propose an IMP-GCN model which exploits high-order neighbors from the same subgraph, in which the user nodes share more similar interests than those in other subgraphs. It is proved to be effective on alleviating the over-smoothing problem.
- We design a subgraph generation module to group users and generate subgraphs from the user-item bipartite graph by considering users features and graph structure information.
- We conduct empirical studies on three benchmark datasets to evaluate the proposed IPM-GCN model. Results show that IPM-GCN can gain improvement by stacking more layers and learn better user/item embeddings, and thus outperforms the SOTA GCN-based recommendation models with a large margin.

The rest of this chapter is organized as follows: Section 2.2 provides a comprehensive review of related work. Section 2.3 introduces a recap of the GCN-based recommendation model. Section 2.4 details the components of the proposed model. Experimental results are presented in Section 2.5. Finally, Section 2.6 concludes the chapter.

2.2 Related Work

2.2.1 Model-based Colloborative Filtering

As one of the most important information retrieval techniques, recommendation has made tremendous progress in past decades. Among various recommendation approaches, the model-based collaborative filtering (CF) [113, 76, 112, 135, 56, 22, 136, 20, 54] achieves great success and becomes the mainstream recommendation technique.

---

[1] In experiments, we found that by stacking 7 layers, a user node almost reaches all the other users in three different datasets. Therefore, no more gain after stacking 7 layers.

[2] https://github.com/liufancs/IMP_GCN.

Earlier research efforts mainly focus on the shallow models, such as BPR [112], matrix factorization (MF) [76, 60, 112, 56, 132, 85, 169]. Their success motivates the development of various variants via leveraging additional information (*e.g.*, review [103], image [51], knowledge graph [129, 128, 135]) to deal with different tasks (*e.g.*, context-aware [96], session-based [97]). With the rise of deep learning, it has also been widely applied in recommendation and exhibits great potential by enhancing the user/item embedding learning, promoting another peak development of recommendation technique. Many DL-based recommendation models have been proposed, such as NeuMF [56], Wide&Deep [19], and achieved better performance over traditional models. For example, NeuMF [56] models the nonlinear interactions between users and items using nonlinear neural networks as the interaction function.

CF technique learns user and item embeddings by reconstructing the user-item interaction matrix. After obtaining the embedding vectors for both the user and the item, a preferred score is calculated between them using the dot product [112, 76]. However, the dot product has an intrinsic problem in that it does not satisfy the triangle inequality [59, 86]. To address this problem, Hsieh et al. [59] proposed a collaborative metric filtering (CML) method that learns user preferences by minimizing the distance between the vectors of positive interactions between the user and the item. Due to the intricate nature of user-item interactions, deep learning techniques are also employed as the interaction prediction function in recommendation systems [56].

Despite great success, the above-mentioned methods are still unable to obtain the optimal embeddings for recommendation. The reason is that the CF signals are only implicitly captured, so the transitivity property of behavior similarity could be captured [136]. A common solution is to explicitly leverage the graph structure to assist the user and item representation learning. In the following, we will focus on discussing Graph-based recommendation methods.

2.2.2 Graph-based Recommendation Models

Another research line is graph-based recommendation, which can explicitly exploit high-order proximity between users and items. Early approaches infer indirect preference by random walks in the graph to provide recommendation [46, 27, 42]. The recently proposed approaches exploit the user-item bipartite graph to enrich the user-item interactions [156, 158] and explore other types of collaborative relations, such as user-user and item-item similar ties [9, 158]. For example, HOP-Rec [156] uses random sample positive user-item interactions to enrich the training data by using random walks. WalkRanker [158] and CSE [9] performs random walks to explore the high-order proximity in user-user and item-item relations. As those methods rely on random walks to sample new interactions for model training, their performance heavily depends on the quality of generated interactions by random walks. As a result, these methods need carefully selection and tuning effects.

In recent years, Graph Convolution Networks (GCNs) have attracted increasing attention in recommendation due to the powerful capability on representation learning from non-Euclidean structure [5, 136, 157, 38, 140, 135, 54, 163, 144, 143, 88, 36]. And then, many GCN-based recommendation models have been developed. For example, GC-MC [5] employs one convolution layer to exploit the direct connections between users and items; PinSage [157] combines random walks with multiple graph convolution layers on the item-item graph for Pinterest image recommendation; MEIRec [36] utilizes metapath-guided neighbors to exploit rich structure information for intent recommendation; NGCF [136] exploits high-order proximity by propagating embeddings on the user-item interaction graph; instead of implicitly capturing the high-order connectivity through the propagation embedding. Inspired by the study of simplifying GCN [145], researchers also introspect the complex design in GCN-based recommendation models. He et al. [54] pointed out that the two common designs feature transformation and nonlinear activation have no positive effects on the final performance, and proposed LightGCN which substantially improves the performance over NDCG. Meanwhile, Chen et al. [16] also proposed to remove the nonlinearity in the network and introduced a residual network to alleviate the over-smoothing problem in existing GCN-based recommendation models.

In this chapter, we move a step further on this research line. We claim that indiscriminately exploiting the high-order neighboring nodes is also an important reason for the over-smoothing problem for the GCN-based recommendation model. A typical example is that two users with contradictory interests can also be connected via a k-order path in the user-item interaction graph. To tackle the problem, we propose an interest-aware message-passing strategy to make the embedding propagation only happened inside a subgraph with similar interests.

2.3 Recap

Let $\boldsymbol{A} \in \mathbb{R}^{N \times M}$ be the user-item interaction matrix, where N and M indicate the number of users and items, respectively. An nonzero entry $a_{ui} \in \boldsymbol{A}$ indicates that user $u \in \mathcal{U}$ has interacted with item $i \in \mathcal{I}$ before; otherwise, the entry is zero. A user-item bipartite graph $\mathcal{G} = (\mathcal{W}, \mathcal{E})$ can be constructed based on the interaction matrix, where the node set $\mathcal{W}$ consists of the two types of user nodes and item nodes and $\mathcal{E}$ represents for the set of edges. For a nonzero a_{ui}, there is an edge between the user u and item i. The above information is taken as the input of GCN model to learn the user and item representations by iteratively aggregating features from neighboring nodes in the bipartite graph.

Here we take LightGCN as an example to describe the GCN-based recommendation model, because it achieves the state-of-the-art performance with a very light design. Our model is also developed based on its design.[3] Let $\boldsymbol{e}_u^{(0)}$ denote the ID em-

[3] Note that although LR-GCN was inspired by a different motivation, its final formulation is almost the same as LightGCN.

bedding of user u and $\boldsymbol{e}_i^{(0)}$ denote the ID embedding of item i, the graph convolution operation in LightGCN is described as follows:

$$\begin{aligned}\boldsymbol{e}_u^{(k)} &= \sum_{i\in\mathcal{N}_u} \frac{1}{\sqrt{|\mathcal{N}_u|}\sqrt{|\mathcal{N}_i|}}\boldsymbol{e}_i^{(k-1)},\\ \boldsymbol{e}_i^{(k)} &= \sum_{u\in\mathcal{N}_i} \frac{1}{\sqrt{|\mathcal{N}_i|}\sqrt{|\mathcal{N}_u|}}\boldsymbol{e}_u^{(k-1)},\end{aligned} \tag{2.1}$$

where $\boldsymbol{e}_u^{(k)}$ and $\boldsymbol{e}_i^{(k)}$ represent the embeddings of the user u and item i after k layers propagation, respectively; $\mathcal{N}_u$ denotes the set of items that interact with user u, and $\mathcal{N}_i$ denotes the set of users that interact with item i; $\frac{1}{\sqrt{|\mathcal{N}_u|}\sqrt{|\mathcal{N}_i|}}$ is symmetric normalization terms, which can avoid the scale of embeddings increasing with graph convolution operations [71]. After K layers graph convolution, the final embeddings of a user u and an item i are the combination of their embeddings obtained at each layer in LightGCN:

$$\boldsymbol{e}_u = \sum_{k=0}^{K} \alpha_k \boldsymbol{e}_u^{(k)}; \boldsymbol{e}_i = \sum_{k=0}^{K} \alpha_k \boldsymbol{e}_i^{(k)}, \tag{2.2}$$

where $\alpha_k \geq 0$ is a hyper-parameter assigned to the k-th layer. It denotes the importance of this layer in constituting the final embedding. From Eq. 2.2, it is expected that after iteratively aggregating features from higher-order neighbors, the nodes will fail to preserve their own distinct features and their embeddings become more and more similar, leading to the over-smoothing problem. Besides, it does not distinguish the heterogeneous features of high-order nodes in the aggregation process. The noisy information from high-order neighbors could hurt the embedding learning. For example, the embeddings of users with no common interests or even contradictory interests in the high-order neighbors are aggregated to learn a target user's embedding via the graph convolution operation.

Fig. 2.1 shows the average coverage ratio of *the number of nodes that a target node reaches in the propagation by stacking different numbers of layers* to *all the nodes in the graph*. It can be seen that after 6- or 7-layer graph convolution, a node can almost receive information from all the other nodes in embedding propagation. Therefore, by aggregating information from all the connected high-order neighbors, it is unavoidable that the node embeddings become homogeneous in the current GCN-based models after stacking more layers, especially for the densely connected ones, whose embeddings will become more and more similar. In the recommendation scenario, this means the uniqueness of users will be neglected in deep structure.

Actually, current GCN-based recommendation models achieve their peak performance at most 3 or 4 layers [54, 145]. Besides the over-smoothing effect, we deem that a node also takes noisy or negative information in the embedding propagation process, which hurts the final performance. This is because a user's interests often span a range of items. Different users can have very different interests or even exhibit contradictory attitudes to some items. Without distinguishing those users, the

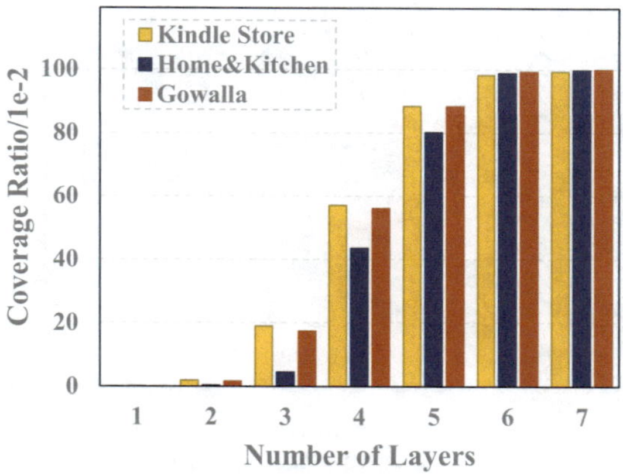

Fig. 2.1: The average ratio of nodes involved in different layers of graph convolution on three datasets.

embedding propagation may perform among users with very different interests to learn their embeddings in the graph convolution operation. To avoid the situation and alleviate the over-smoothing problem, it is important to group users with similar interests (and their interacted items) into subgraphs and constrain the embedding propagation to operate inside the subgraph. To achieve the goal we propose the interest-aware message-passing GCN model.

2.4 IMP-GCN Model

2.4.1 Interest-aware Message-passing Strategy

With constructing subgraphs, we would like that all the information propagated in a subgraph can contribute to the embedding learning of all the nodes in this subgraph. In other words, we aim to exclude the negative information propagation in the graph convolution operation using subgraphs. To achieve the goal, we rely on user nodes to form subgraphs in the user-item bipartite graph. The general idea is that users with more similar interests are grouped into a subgraph, and the items which directly linked to those users also belong to this subgraph. Therefore, each user only belongs to one subgraph, and an item can be associated with multiple subgraph. Let G_s with $s \in \{1, \cdots, N_s\}$ denotes a subgraph, where N_s is the number of subgraphs. In the next, we introduce the graph convolution operation in our model.

Because the direct interactions between users and items provide the most important and reliable information of user interests, in the first-order propagation, all the first-order neighbors are involved in the graph convolution operation. Let $\boldsymbol{e}_u^{(0)}$ and $\boldsymbol{e}_i^{(0)}$ denote the ID embeddings of user u and item i, respectively. The first-order

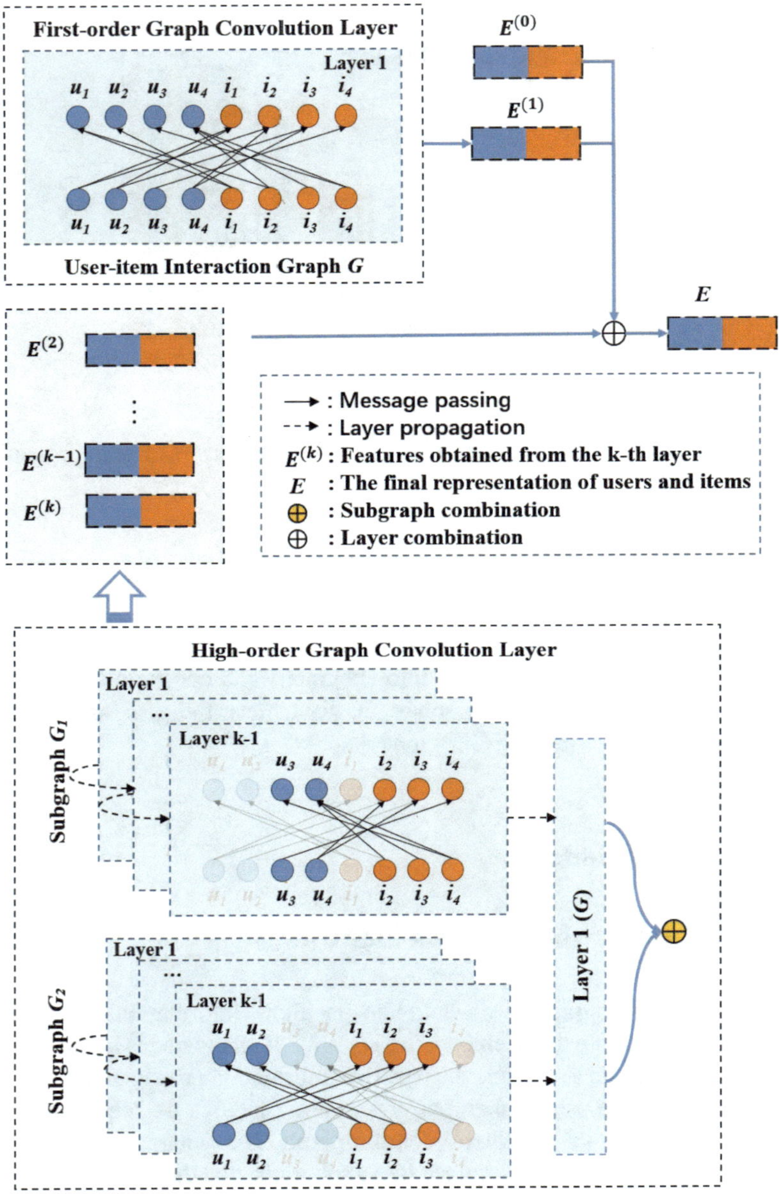

Fig. 2.2: An overview of our IMP-GCN model with two subgraphs as illustration. In IMP-GCN, the first-order propagation operates on whole graph, and high-order propagation operates inside the subgraphs.

graph convolution is:

$$\begin{aligned} \boldsymbol{e_u^{(1)}} &= \sum_{i \in \mathcal{N}_u} \frac{1}{\sqrt{|\mathcal{N}_u|}\sqrt{|\mathcal{N}_i|}} \boldsymbol{e_i^{(0)}}, \\ \boldsymbol{e_i^{(1)}} &= \sum_{u \in \mathcal{N}_i} \frac{1}{\sqrt{|\mathcal{N}_i|}\sqrt{|\mathcal{N}_u|}} \boldsymbol{e_u^{(0)}}, \end{aligned} \tag{2.3}$$

where $\boldsymbol{e_u^{(1)}}$ and $\boldsymbol{e_i^{(1)}}$ represent the first layer embeddings of the target user u and item i, respectively.

For the high-order graph convolution, to avoid introducing noisy information, a node in a subgraph can only exploit the information from its neighbor nodes in this subgraph. Because the items interacted by a user all belong to the subgraph of this user, the user can still receive information from all the linked items. However, for an item node, its direct user neighbors can be distributed in different subgraphs. To learn the embeddings of an item i, for each subgraph G_s it belongs to, we learn an embedding for this item. Let $\boldsymbol{e_{is}^{(k)}}$ denotes the embedding of item i in subgraph s after k layers graph convolution, the high-order propagation in IMP-GCN is defined as:

$$\begin{aligned} \boldsymbol{e_u^{(k+1)}} &= \sum_{is \in \mathcal{N}_u} \frac{1}{\sqrt{|\mathcal{N}_u|}\sqrt{|\mathcal{N}_i|}} \boldsymbol{e_{is}^{(k)}}, \\ \boldsymbol{e_{is}^{(k+1)}} &= \sum_{u \in \mathcal{N}_i^s} \frac{1}{\sqrt{|\mathcal{N}_i|}\sqrt{|\mathcal{N}_u|}} \boldsymbol{e_u^{(k)}}. \end{aligned} \tag{2.4}$$

In this way, we guarantee that the embedding of a node learned in a subgraph only contributes to the embedding learning of other nodes in this subgraph. This can avoid the noisy information propagated from unrelated nodes. $\boldsymbol{e_{is}^{(\cdot)}}$ can be regarded as the features learned from the users with a similar interest in the subgraph G_s. This make senses since users with similar interests often prefer the same feature of an item. The final representation of an item i after k layers graph convolution is a combination of its embeddings learned in different subgraphs, *i.e.*,

$$\boldsymbol{e_i^{(k)}} = \sum_{s \in \mathcal{S}} \boldsymbol{e_{is}^{(k)}}, \tag{2.5}$$

where S is the subgraph set that item i belongs to.

2.4.2 Layer Combination and Prediction

We combine the embeddings obtained at each layer to form the final representation of user u and item i as Eq. 2.2. Similar to LightGCN, α_k is set uniformly as $1/(K+1)$ [54].

With the learned embeddings of users (*i.e.*, $\boldsymbol{e}_u$) and items $\boldsymbol{e}_i$, given a user u and a target item i, the preference of the user to the item is computed by inner product:

$$\hat{r}_{uv} = \boldsymbol{e_u^T} \boldsymbol{e_i}. \tag{2.6}$$

Notice that other interaction functions can be also applied, such as Euclidean distance. Because the main focus of this work is to study the effects of distinguishing user interests in the graph convolution in the GCN-based recommendation model, we adopt the inner product as previous work [156, 9, 136] for fair comparisons in the empirical studies.

2.4.3 Matrix-form propagation rule

We implement our algorithm with the matrix form propagation rule (see [136] for more details), by which we can simultaneously update the representations of all users and items in a rather efficient way. It is a commonly used approach to make graph convolution network feasible for large-scale graph [136, 109]. Let $\boldsymbol{E}^{(0)}$ be the representations matrix for users ID and items ID; $\boldsymbol{E}^{(\boldsymbol{k})}$ represents the representation of users and items at the k-th layer. Similarly, $\boldsymbol{E}_{\boldsymbol{s}}^{(\boldsymbol{k})}$ is defined as the representation of users and items at the k-th layer in subgraph G_s. As shown in Fig. 2.2, the first layer embedding propagation in our model can be described as follows:

$$\boldsymbol{E}^{(\mathbf{1})} = \mathcal{L}\boldsymbol{E}^{(\mathbf{0})}, \tag{2.7}$$

where $\mathcal{L}$ is the Laplacian matrix for the user-item interaction graph.

As we involve the subgraphs in high-order graph convolution layers,the embeddings propagation on subgraphs is formulated as follows:

$$\boldsymbol{E}_{\boldsymbol{s}}^{(\boldsymbol{k}-\mathbf{1})} = \mathcal{L}_s\boldsymbol{E}_{\boldsymbol{s}}^{(\boldsymbol{k}-\mathbf{2})}, \tag{2.8}$$

where $k \geqslant 2$; $\mathcal{L}_s$ represent the Laplacian matrix for the subgraph G_s. And then, the $(k-1)$-th layer embeddings are propagated on the user-item graph and obtained the embeddings in the k-th layer:

$$\boldsymbol{E}_{\boldsymbol{s}}^{(\boldsymbol{k})} = \mathcal{L}\boldsymbol{E}_{\boldsymbol{s}}{}^{(\boldsymbol{k}-\mathbf{1})}. \tag{2.9}$$

We aggregate all the k-th layer embeddings involved different subgraphs to formulate the final k-th layer embeddings:

$$\boldsymbol{E}^{(\boldsymbol{k})} = \sum_{s \in G_s} \boldsymbol{E}_{\boldsymbol{s}}^{(\boldsymbol{k})}. \tag{2.10}$$

Lastly, we combine all the layers' embeddings and get the final representations of users and items, this formulation keeps consistent with it in LightGCN [54]:

$$\boldsymbol{E} = \alpha_0\boldsymbol{E}^{(\mathbf{0})} + \alpha_1\boldsymbol{E}^{(\mathbf{1})} + \cdots + \alpha_K\boldsymbol{E}^{(\boldsymbol{K})} \tag{2.11}$$

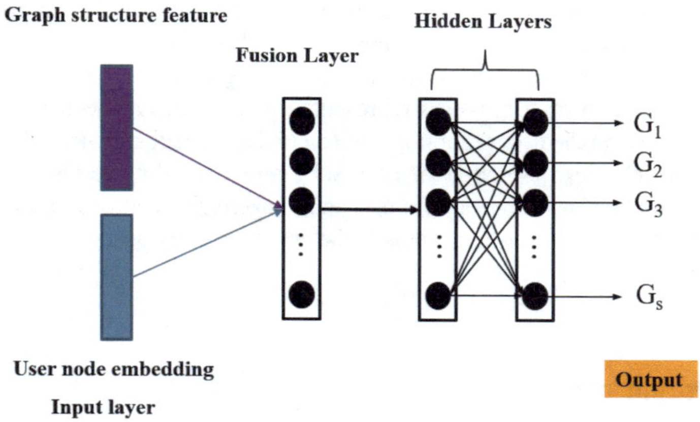

Fig. 2.3: User classifying. The position of the maximum value in the output indicates the group to which the user belongs.

2.4.4 Subgraph Generation Module

In this section, we introduce our proposed subgraph generation module which is designed to construct the subgraphs G_s with $s \in \{1, \cdots, N_s\}$ from a given input graph $\mathcal{G}$. Remind that the subgraphs are used to group users with common interests in our model. As shown in Fig. 2.3, we formulate the user grouping as a classification task [62], *i.e.*, each user is classified to a group. Specifically, each user is represented by a feature vector, which is a fusion of the graph structure and the ID embedding:

$$\boldsymbol{F_u} = \sigma(\boldsymbol{W_1}(\boldsymbol{e}_u^{(0)} + \boldsymbol{e}_u^{(1)}) + \boldsymbol{b_1}), \tag{2.12}$$

where $\boldsymbol{F_u}$ is the obtained user feature via feature fusion. $\boldsymbol{e}_u^{(0)}$ is the embedding of user ID and $\boldsymbol{e}_u^{(1)}$ is the feature obtained by aggregating local neighbor in the graph (*i.e.*, the user embedding after the first layer propagation.). $\boldsymbol{W_1} \in R^{d\times d}$ and $\boldsymbol{b_1} \in R^{1\times d}$ are respectively the weight matrix and bias vector of the fusion method. σ is the activation function. LeakyReLU [100] is adopted, because it can encode both positive and small negative signals. To classify the users into different subgraphs, we cast the obtained user feature to a prediction vector with a 2-layer neural networks:

$$\begin{aligned} \boldsymbol{U_h} &= \sigma(\boldsymbol{W_2 F_u} + \boldsymbol{b_2}), \\ \boldsymbol{U_o} &= \boldsymbol{W_3 U_h} + \boldsymbol{b_3}, \end{aligned} \tag{2.13}$$

where $\boldsymbol{U_o}$ is the prediction vector. The position of maximum value in U_o represents which group/subgraph the user belongs to. $\boldsymbol{W_2} \in R^{d\times d}$, $\boldsymbol{W_3} \in R^{d\times N_s}$ and $\boldsymbol{b_2} \in R^{1\times d}$, $\boldsymbol{b_3} \in R^{1\times N_s}$ are respectively the weight matrices and bias vectors of the two layers. The dimension of the prediction vector dimensions is the same as the number of subgraphs, which is a pre-selected hyper-parameter. Note that it is an unsupervised method to classify users into different groups and thus does not need ground-truth label. For users with similar embeddings, Eq. 2.13 will generate similar prediction vector, namely, they will be classified into the same group. The subgraph generation aims to construct a matrix, which represents the user-item adjacency relation in a subgraph based on the user grouping results and the Laplacian matrix of the original user-item graph. For the matrix of each subgraph, according to the obtained user group information, we filter out the user-item adjacency relations in the Laplacian matrix of the original user-item graph if the corresponding users are not in the user group.

2.4.5 Optimization

In this work, we target at the top-n recommendation, which aims to recommend a set of n top-ranked items matching the target user's preference. Compared to rating prediction, this is a more practical task in real commercial systems [112]. Similar to other rank-oriented recommendation works [156, 136], we adopt the pairwise learning method for optimization. To perform the pairwise learning, it needs to constructs a triplet of $\{u, i^+, i^-\}$, with *an observed interaction between* u *and* i^+ and *an unobserved interaction between* u *and* i^-. This method assumes that a positive item (*i.e.*, i^+) should rank higher than an negative item (*i.e.*, i^-). The objective function is formulated as:

$$\arg\min \sum_{(\mathbf{u},\mathbf{i}^+,\mathbf{i}^-)\in\mathcal{O}} -\ln\phi(\hat{r}_{ui^+} - \hat{r}_{ui^-}) + \lambda\,\|\Theta\|_2^2 \tag{2.14}$$

where $\mathcal{O} = \{(u, i^+, i^-)|(u, i^-) \in \mathcal{R}^+, (u, i^-) \in \mathcal{R}^-\}$ denotes the training set; $\mathcal{R}^+$ indicates the observed interactions between user u and i^+ in the training dataset, and $\mathcal{R}^-$ is the sampled unobserved interaction set. λ and Θ represent the regularization weight and the parameters of the model, respectively. The L_2 regularization is used to prevent overfitting.

The mini-batch Adam [70] is adopted to optimize the prediction model and update the model parameters. Specifically, for a batch of randomly sampled triples $(u, v^+, v^-) \in (O)$, the representation of those users and items are first learned by the propagation rules and then the model parameters are updated by using the gradients of the loss function.

2.5 Experiments

To evaluate the effectiveness of our proposed IMP-GCN, we conducted extensive experiments on four public datasets. In particular, we primarily answer the following questions.

Table 2.1: Basic statistics of the experimental datasets.

Dataset	#user	#item	#interactions	sparsity
Kindle Store	68,223	61,934	982,618	99.98%
Home&Kitchen	66,519	28,237	551,681	99.97%
Gowalla	29,858	40,981	1,027,370	99.92%

RQ1: How do the number of graph convolution layers and subgraphs impact IMP-GCN?

RQ2: Does our proposed IMP-GCN outperform state-of-the-art CF baselines in the top-n recommendation task?

RQ3: How do the components of IMP-GCN contribute to enhancing recommendation performance?

In the following, we present the experimental setup and answer the above questions sequentially based on the experimental results.

2.5.1 Experimental Setup

Data Description

To evaluate the effectiveness of IMP-GCN, we conducted experiments on three benchmark datasets: Amazon-Kindle Store, Amazon-Home&Kitchen and Gowalla.

The first two datasets are from the public Amazon review dataset[4], which has been widely used for recommendation evaluation in previous studies [103]. The Amazon Review Dataset is a large collection of product reviews from Amazon, which includes user-generated feedback on various products across different categories. It is widely used in data science, machine learning, and natural language processing (NLP) tasks, particularly for sentiment analysis, recommendation systems, and text mining. This public dataset contains user interactions (review, rating, helpfulness votes, etc.) on items and the item metadata (descriptions, attributes, images, etc.) on 24 product categories.

The third dataset is a check-in dataset collected from Gowalla, where users share their locations by checking-in. We followed the general setting in recommendation to filter users and items with few interactions. For all the datasets, we used the 10-core settings, *i.e.*, retaining users and items with at least 10 interactions. The statistics of three datasets are shown in Table 2.1. As we can see, the datasets are of different sizes and sparsity levels, which are useful for analyzing the performance of our method and the competitors in different situations.

[4] http://jmcauley.ucsd.edu/data/amazon.

For each datasets, we randomly split it into training, validation, and testing set with the ratio 80:10:10 for each user. The observed user-item interactions were treated as positive instances. For the methods which adopt the pairwise learning strategy, we randomly sample a negative instance, that the user did not consume before, to pair with each positive instance.

Evaluation Metrics

For each user in the test set, we treat all the items that the user did not interact with as negative items. Two widely used evaluation metrics for top-n recommendation are adopted in our evaluation: **Recall** and **NDCG** (Normalized Discounted Cumulative Gain) [52].

Recall measures the proportion of relevant items successfully retrieved by the recommendation algorithm. It is defined as:

$$\text{Recall} = \frac{|\text{Rel}(u) \cap \text{Rec}(u, n)|}{|\text{Rel}(u)|}, \tag{2.15}$$

where $\text{Rel}(u)$ is the set of relevant items for user u, and $\text{Rec}(u, n)$ is the set of top-n recommended items for user u.

NDCG is a commonly used metric in information retrieval to evaluate the quality of ranked lists. It considers the relevance of items and their positions in the ranking, emphasizing higher relevance at higher ranks. The Discounted Cumulative Gain (DCG) is given by:

$$\text{DCG}_p = \sum_{i=1}^{p} \frac{2^{\text{rel}_i} - 1}{\log_2(i+1)},$$

where:

- rel_i is the relevance score of the item at position i,
- p is the rank position until which DCG is computed.

The Ideal Discounted Cumulative Gain (IDCG) is the maximum possible DCG, computed using an ideal ranking of items:

$$\text{IDCG}_p = \sum_{i=1}^{p} \frac{2^{\text{rel}_i^{\text{ideal}}} - 1}{\log_2(i+1)},$$

where $\text{rel}_i^{\text{ideal}}$ represents the relevance scores in the ideal ranking. The Normalized Discounted Cumulative Gain is then calculated as:

$$\text{NDCG}_p = \frac{\text{DCG}_p}{\text{IDCG}_p}.$$

NDCG scores range from 0 to 1, where 1 indicates a perfect ranking.

For each metric, the performance is computed based on the top 20 results. Notice that the reported results are the average values across all the testing users.

Experimental Settings

We implemented our model with Tensorflow [5] and carefully tuned the key parameters. The embedding size is fixed to 64 for all models and the embedding parameters are initialized with the Xavier method [149]. We optimized our method with Adam [70] and used the default learning rate of 0.001 and default mini-batch size of 1024 (on gowalla, we increased the mini-batch size to 2048 for speed). The L_2 regularization coefficient λ is searched in the range of $\{1e^{-6}, 1e^{-5}, \cdots, 1e^{-2}\}$. The early stopping and validation strategies are kept the same as those in LightGCN.

2.5.2 Study of IMP-GCN (RQ1)

In this section, we first evaluated the performance of our IPM-GCN model when stacking different layers in graph convolution. This is to examine whether our interest-aware message-passing strategy can alleviate the over-smoothing problem. In the next, we study the effects of the subgraph numbers on the performance of our model.

Effect of Layer Numbers

To investigate the effectiveness of IMP-GCN in deeper structure, we increased the model depth and performed detailed comparison with LightGCN. Since the adopted message-passing strategy is the same as LightGCN in the first-order convolution layer, we increased the layer number from 2 to 7. The experimental results are shown in Fig. 2.4, in which IMP-GCN_2, IMP-GCN_3 and IMP-GCN_4 indicate the model with 2, 3, and 4 subgraphs, respectively.

Firstly, the proposed IMP-GCN outperforms LightGCN consistently when stacking more than 2 or 3 layers over both datasets. This indicates that our model can learn better embeddings by the interest-aware message-passing strategy. Secondly, the peak performance of LightGCN is obtained when stacking 3 or 4 layers, and increasing more layers will cause dramatic performance degradation, indicating it suffers from the over-smoothing problem in a deep structure. In contrast, IMP-GCN continues to achieve better performance with deeper structure (notice that when stacking more than 7 layers, a node already aggregates information from almost all the nodes, see Fig. 2.1. The results demonstrate the capability of our model on alleviating the over-smoothing problem. Moreover, it also 1) justifies our claim that exploiting information from all nodes indiscriminately causes the over-smoothing in GCN-based recommendation model, and 2) validates the effectiveness of our sub-

[5] https://www.tensorflow.org.

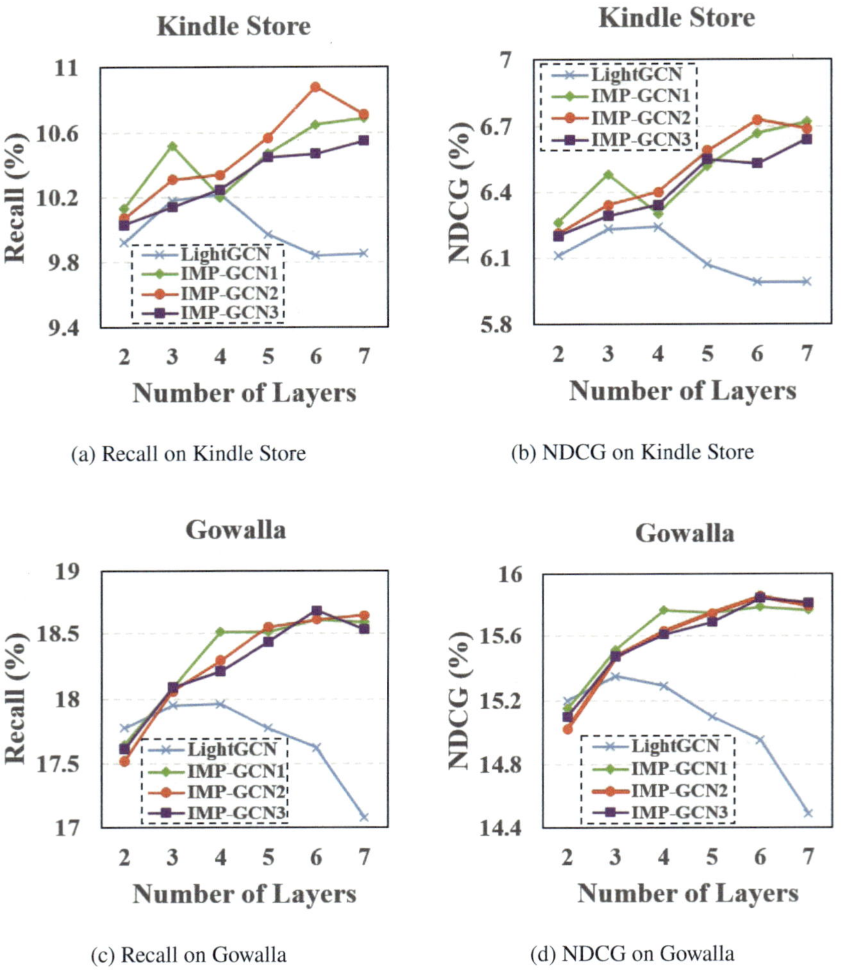

(a) Recall on Kindle Store

(b) NDCG on Kindle Store

(c) Recall on Gowalla

(d) NDCG on Gowalla

Fig. 2.4: Results Comparison between IMP-GCN and LightGCN at different layers on Kindle Store and Gowalla. IMP-GCN$_2$, IMP-GCN$_3$, and IMP-GCN$_4$ represent IMP-GCN with 2, 3, and 4 subgraphs, respectively.

graph generation algorithm on classifying users with common interests.

Effect of Subgraph

The performance of IPM-GCN with different numbers (*i.e.*, $\{2, 3, 4\}$) of subgraphs can also be observed in Fig. 2.4. From the results, we can see that the (1) IMP-GCN$_2$ with 2 subgraphs can obtain the best results when stacking no more than 3 layers. This is because a node in the subgraphs of IMP-GCN$_2$ can reach more nodes in

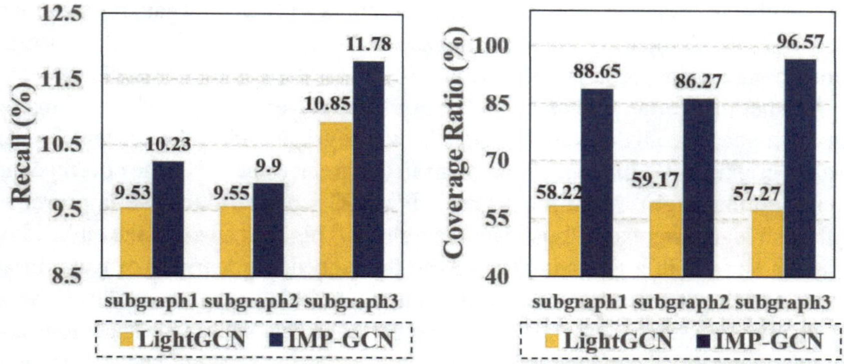

(a) Recall on Kindle Store (b) Coverage Ratio on Kindle Store

Fig. 2.5: Statistics of Recall and Coverage Ratio on Kindle Store in three subgraphs.

short distance than the on in IMP-GCN$_3$ or IMP-GCN$_4$in the embedding propagation operation. (2) When stacking more than 3 layers, IMP-GCN$_3$ performs the best. After 3 layers graph convolution, the number of involved nodes increasing sharply in embedding propagation (see the examples in Fig. 2.1). On average, each node in IMP-GCN$_2$ should reach more nodes that the one in IMP-GCN$_3$ and IMP-GCN$_4$, however, the performance improvement of IMP-GCN$_2$ is smaller or even negative (on th Kindle Stores) than that of IMP-GCN$_3$ and IMP-GCN$_4$. This indicates that there is still noisy information in embedding propagation by discriminating user interests in a coarse-level (*i.e.*, 2 subgraphs), negatively impacting the performance. Note that IMP-GCN$_3$ can still benefit from high-order neighbors. (3) With more subgraphs, on the one hand, IMP-GCN$_4$ can distinguish users with similar interests in a finer level and thus can better distill information from high-order neighbors; on the other hand, it also cuts more connections to other nodes, especially the ones in short distance which provide more valuable information in embedding learning. As a result, when stacking more layers, its performance is only comparable to that of IMP-GCN$_2$. Therefore, there is a trade-off on selecting the number of subgraphs. We further studied the effects of subgraphs by analyzing the average coverage ratio of each node and the corresponding performance based on the LightGCN and our IPM-GCN model. Due to the space limitation, we only provide the results on Kindle Store and omit the performance $w.r.t$ ndcg which has the similar trend as recall. In this experiment, we used the LightGCN with 4 layers and IPM-GCN with 3 subgraphs[6] and 6 layers, which are their optimal setting on Kindle Store. The average recall and average cover ratio of each user in a subgraph based on LightGCN and IPM-GCN are shown in Fig. 2.5(a) and Fig. 2.5(b), respectively. Notably, by grouping users with similar interest in subgraphs to make information only propagate inside subgraphs, IPM-GCN can benefit from more layers of graph convolution and distill

[6] Number of users in the three groups G_1, G_2, G_3 are $3,971, 3,584, 6,801$, respectively.

positive information from high-order neighors. In contrast, LightGCN is limited by the negative information from high-order neighbors and can only gain improvements over 4 layers. Comparing the performance of different subgraphs, we can see that with a higher coverage ratio, the performance of IPM-GCN increases clearly.

Another interesting finding is that, by stacking 6 layers, a user node in a subgraph almost connects to all the other nodes in the whole graph. This indicates that the users in a subgraph almost interact all the items in the graph (otherwise, the coverage ratio cannot be that high). More importantly, IPM-GCN can still achieve improvement with such high coverage without over-smoothing. This indicates that the embeddings of items learned in a graph contributes to the embedding learning of users in this graph, and the distilled information in a subgraph during graph convolution is useful for the embedding learning for all the nodes in this subgraph. It demonstrates the effectiveness of our interest-aware message-passing strategy and the subgraph generation algorithm.

2.5.3 Comparison with SOTA Methods (RQ2)

Baselines

To demonstrate the effectiveness, we compared our proposed method with several recently proposed competitive methods, including

- **NeuMF [56]**: It is a state-of-the-art neural collaborative filtering method. This method uses multiple hidden layers above the element-wise and concatenation of user and item embeddings to capture their non-linear feature interactions [7].
- **HOP-Rec [156]:** This method exploits the high-order user-item interactions by random walks to enrich the original training data. In experiments, we used the codes released by the authors [8].
- **CSE [9]:** This recently proposed graph-based model also exploits the high-order proximity in the user-item bipartite graph. Different from HOP-Rec, this method explores the user-user and item-item relations by random walks to improve the performance. We used the codes released by the authors (the same link as HOP-Rec).
- **GCMC [5]:** This method applies the GCN techniques on user-item bipartite graph and employs one convolutional layer to exploit the direct connections between users and items [9].
- **NGCF [136]**: This method explicitly encodes the collaborative signal in the form of high-order connectivities by performing embedding propagation in the user-item bipartite graph [10].

---

[7] https://github.com/hexiangnan/neural_collaborative_filtering

[8] https://github.com/cnclabs/smore.

[9] https://github.com/riannevdberg/gc-mc

[10] https://github.com/xiangwang1223/neural_graph_collaborative_filtering

Table 2.2: Performance of our model and the competitors over three datasets. Noticed that the values are reported by percentage with '%' omitted.

Datasets	Kindle Store		Home&Kitchen		Gowalla	
Metrics	Recall	NDCG	Recall	NDCG	Recall	NDCG
NeuMF	4.96	2.06	1.34	0.62	12.96	11.21
CSE	7.65	4.54	1.93	0.91	13.85	11.51
HOP-Rec	7.96	4.58	1.98	0.94	14.11	12.70
GCMC	7.93	4.55	1.42	0.64	14.03	11.68
NGCF	8.25	5.09	2.14	0.96	15.62	13.35
LightGCN	**10.22**	**6.24**	**3.03**	**1.39**	**17.96**	**15.29**
IMP-GCN	**10.88***	**6.73***	**3.22***	**1.49***	**18.69***	**15.85***
Improv.	6.46%	7.85%	6.27%	7.19%	4.07%	3.66%

The symbol * denotes that the improvement is significant with $p - value < 0.05$ based on a two-tailed paired t-test.

- **LightGCN [54]**: It is an simplified version of NGCF by removing the feature transformation and nonlinear activation module. It makes GCN-based methods more concise and appropriate for recommendation and achieves the state-of-the-art performance [11].

For fair comparisons, all the methods are optimized by the same pairwise learning strategy. We put great efforts into tuning these methods based on the validation dataset and reported their best performance.

Overall Comparison

Table 2.2 shows the performance comparison results. The best and second best results were highlighted in bold. From the results, we had following observations.

The performance of NeuMF is relatively poor as it not explicitly leverages the high-order connectivities between users and items, resulting in suboptimal performance. For the graph-based methods, CSE makes use of the implicit associates of user-user and item-item similarities via high-order neighborhood proximity by performing random walks on the user-item interaction graph. GCMC obtains better performance over CSE, demonstrating the advantages of GCN-based approaches, which can exploit graph structure information. However, it does not perform well on $Home\&Kitchen$ because the useful information in neighbors cannot be efficiently aggregated. Hop-Rec outperforms the above methods on the three datasets, because it samples user-item interactions from high-order neighbors to enrich the training data. NGCF achieves consistent much better performance over the above baselines. This is because it adopts the GCN techniques to explicitly and directly exploit the high-order connectivities in the embedding. In contrast, the GCMC method only utilizes

[11] https://github.com/kuandeng/LightGCN

Table 2.3: Performance of our model and its variants over three datasets. Noticed that the values are reported by percentage with '%' omitted.

Datasets	Kindle Store		Home&Kitchen		Gowalla	
Metrics	Recall	NDCG	Recall	NDCG	Recall	NDCG
IMP-GCN$_s$	10.57	6.63	3.14	1.43	18.61	15.61
IMP-GCN$_f$	10.19	6.40	2.97	1.31	17.84	15.11
IMP-GCN	**10.88**	**6.73**	**3.22**	**1.49**	**18.69**	**15.85**

the first-order neighbors for representation learning; HOP-Rec and CSE leverage the high-order neighbors to enrich the training data rather than using them in embedding function for direct representation learning. This demonstrates the powerful representation learning capability of GCN and the importance of utilizing high-order information directly in representation learning. Similar to the results reported in [54], LightGCN achieves substantially improvement over NGCF by simplifying it with the removal of two common designs in GCN.

IMP-GCN outperforms all the baselines consistently over all the datasets. In particular, compared to the strongest baseline in terms of NDCG@20, IMP-GCN can reach a relative improvement over LightGCN by 7.85%, 7.19%, 3.66% on $KindleStore$, $Home\&Kitchen$ and $Gowalla$, respectively. The great improvement over LightGCN demonstrates the importance of distinguishing nodes in high-order neighbors in the graph convolution operation, as well as the effectiveness of our proposed interest-aware message-passing strategy.

2.5.4 Ablation Study (RQ3)

In this section, we examined the contribution of different components in our model to the final performance by comparing IMP-GCN with the following two variants:

- **IMP-GCN$_s$**: This variant removes the graph structure information from the subgraph generation module (*i.e.*, removing $\boldsymbol{e}_u^{(1)}$ in Eq. 2.12).
- **IMP-GCN$_f$**: In this variant, the first-order propagation is also performed inside each subgraph (*i.e.*, The equation for $\boldsymbol{e}_i^{(1)}$ in Eq. 2.3 is replaced with $\sum_{s\in\mathcal{S}} \boldsymbol{e}_{is}^{(1)}$).

The results of two variants and IPM-GCN were reported in Table 2.3, in which the best results are highlighted in bold. IMP-GCN outperforms IMP-GCN$_s$ over all the datasets, which indicates the effectiveness of employing graph structure information in subgraph generation module. It is expected that IMP-GCN$_s$ obtains much better performance over IMP-GCN$_f$, because the first-order neighbors (*i.e.*, the interaction between users and items) contributes the direct information for user and embedding in the collaborative filtering process. The results also demonstrate the reasonable design of our IPM-GCN model.

2.6 Conclusion

In this chapter, we argued that exploiting high-order node indiscriminately would introduce negative information into the embedding propagation in the GCN-based recommendation models, causing the performance degradation when stacking more layers. We presented a IMP-GCN model which learns user and item embeddings by performing high-order graph convolution inside subgraphs. The subgraphs are formed by a designed subgraph generation algorithm that groups users with similar interests and their interacted items into the same graph. In IMP-GCN, the embedding of a node learned in a subgraph only contributes to the embedding learning of other nodes in this subgraph. In this way, IMP-GCN can effectively avoid taking the noisy information into the embedding learning. Experiments on large-scale real-world datasets demonstrate that IMP-GCN can gain improvements by stacking more layers to exploit information from higher-order neighbors, and achieve the state-of-the-art performance. The advantages of IMP-GCN indicate the importance of distinguishing high-order neighbors on tackling the over-smoothing problem in GCN models. We believe the insights in this study can shed light on the further development of graph-based recommendation models.

Chapter 3
Cluster-based Graph Collaborative Filtering

3.1 Introduction

In Chapter 2, we explored the method for alleviating the over-smoothing problem in GCN-based recommendation methods. The proposed IMP-GCN [87] maintains the uniqueness of the users, even when additional graph convolution layers are stacked. Building on this, this chapter focuses on capturing users' multiple interests to filter out the noisy information from high-order neighboring nodes, thereby enhancing the robustness of representation learning.

Despite the notable success of GCNs in recommendation systems, a key challenge persists in the information aggregation process of neighboring nodes. Specifically, **noisy information from unreliable neighboring nodes harms the representation of the target nodes** [87, 37]. Most existing GCN-based recommendation models are based on a user-item graph where interactions link users and items. They generally assume that embedding learning could take advantage of collaborative signals from high-order neighbors. Thus, they aggregate the messages passing from neighboring nodes to form the target user/item representation without distinguishing the high-order neighbors. As a result, the noisy information from unreliable users and items is also involved in the representation learning of a target node, resulting in sub-optimal performance. For instance, user u_1 has a strong preference for mystery novels. However, a high-order neighboring user u_2, who is loosely connected to u_1, prefers gardening books. In GCN-based models, this noisy information about gardening books could be delivered to the representation of u_1. Some efforts have been made to enhance the quality of information delivered from the neighbor nodes. For example, IMP-GCN [87] devises an interest-aware strategy to only pass messages from user nodes with common interest and their associated item nodes when performing high-order graph convolution. To capture the adaptive reliability of the interactions between users and items, GTN [37] introduces a graph trend collaborative filtering technique to achieve the desired adaptive smoothness property for the interaction graph. Furthermore, as demonstrated in [106], detecting the users within each community who share similar tastes can improve recommendation accuracy.

F. Liu, L. Nie, *Advancing Recommender Systems with Graph Convolutional Networks*,
https://doi.org/10.1007/978-3-031-85093-6_3

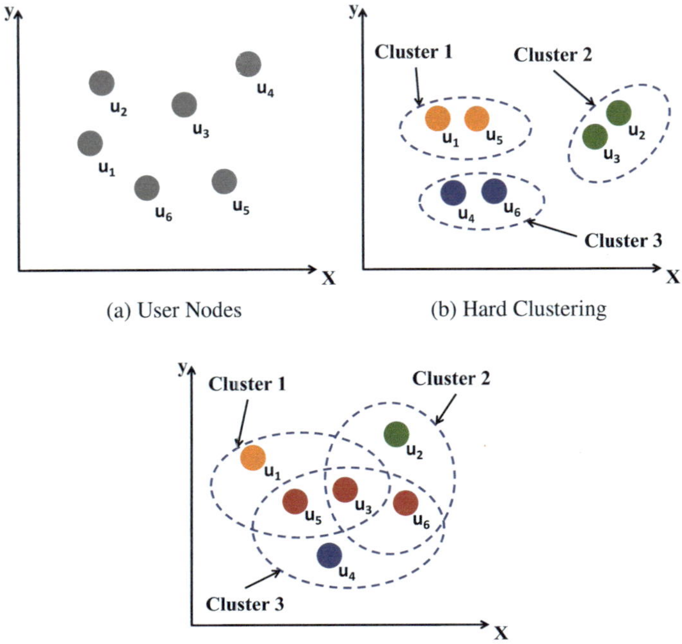

(a) User Nodes (b) Hard Clustering

(c) Soft Clustering

Fig. 3.1: An example of user clustering. (a) A toy example containing 6 user nodes. (b) Hard clustering. Every user node can belong to only one cluster. (c) Soft clustering. Every user node may belong to several clusters with a fractional degree of membership.

Given this challenge, previous works generally ignore high-order connections within the graph, overlooking users' multiple interests in high-order graph convolution operations. Indeed, in [87], the authors demonstrated that users receive more valuable information from other users with common interests and their associated items. As illustrated in Fig. 3.1(a) and (b), to identify the users with common interests, they classify users into different groups using a hard clustering method. However, users often exhibit diverse and overlapping interests across a range of topics or item categories (*e.g.*, movies of different genres). For example, a user might prefer both science fiction and romantic movies. Similarly, a movie is probably categorized into both science fiction and action. The hard clustering fails to capture this overlap, resulting in limited utilization of information. In contrast, with soft clustering, a user could be classified into multiple clusters representing different interests. As shown in Fig. 3.1(c), user u_3 has a common interest with user u_1 in cluster 1 while also sharing an interest with user u_2 within cluster 2. Yet, there is no common interest between u_1 and u_2. Similarly, items can be classified into multiple clusters based on their characteristics. To fully exploit the valuable information from neighboring nodes, it is necessary to jointly consider the multiple interests of each user and the common interest among users for capturing accurate and comprehensive representations of users and items, particularly when performing high-order graph convolution operations.

Motivated by the above consideration, we propose a novel recommendation model termed Cluster-based Graph Collaborative Filtering (ClusterGCF). This model performs high-order graph convolution operations on cluster-specific graphs, which are constructed by capturing the multiple interests of users and identifying the common interests among them. To construct the cluster-specific graphs, we introduce an unsupervised and optimizable soft node clustering approach that classifies all the user and item nodes into multiple clusters by leveraging their embeddings and the graph topology information. Based on the soft node clustering results and the topology of the user-item interaction graph, we assign the nodes with probabilities for different clusters to construct the cluster-specific graphs. Since the embeddings of nodes are propagated within cluster-specific graphs during the graph convolution process, our model can filter out the negative information while capturing more valuable information from high-order neighboring nodes. Besides, the proposed ClusterGCF maintains the uniqueness of node embeddings while stacking more graph convolution layers because it can benefit from higher-order neighboring nodes assigned with different probabilities in cluster-specific graphs. As a result, the over-smoothing problem could be alleviated. Our model's effectiveness was validated through extensive experimentation on four publicly available datasets. Our ClusterGCF results demonstrate superior performance compared to the state-of-the-art GCN-based recommendation methods. This indicates that our model can benefit from higher-order neighbors by classifying nodes into multiple clusters. Furthermore, the results also verified the effectiveness of our proposed soft node clustering approach. We released the codes and involved parameter settings to facilitate others to repeat this work [1].

In summary, the main contributions of this work are as follows:

- We emphasize that GCN-based recommendation methods overlook the users' multiple interests when aggregating information from high-order neighboring nodes. Inspired by this, we propose a ClusterGCF model that performs high-order graph convolution over cluster-specific graphs.
- We develop a soft node clustering approach, which softly classifies the user and item nodes into multiple clusters in an unsupervised and optimizable way. Cluster-specific graphs are constructed based on the clustering results and the topology of the original user-item interaction graph.
- We evaluate the ClusterGCF on four benchmark datasets through empirical studies. Results show that ClusterGCF can gain better user and item representations than SOTA GCN-based recommendation models.

The rest of this chapter is organized as follows: Section 3.2 provides a comprehensive review of related work. Section 3.3 introduces the preliminaries relevant to our proposed model. Section 3.4 details the components of the proposed model.

[1] https://github.com/zhao254014/ClusterGCF.

Experimental results are presented in Section 3.5. Finally, Section 3.6 concludes the chapter.

3.2 related work

In this section, we provide a concise overview of recent developments in the fields of fine-grained user preference modeling and graph convolutional networks. These methods are closely relevant to our work.

3.2.1 Fine-grained User Preference Modeling

Their core idea of Collaborative Filtering (CF) techniques [76, 112, 135, 56, 136, 54, 129, 91] is to learn users' and items' representations ($aka.$ embedding vectors) by exploiting the interaction behavior between them. To improve the recommendation performance, the Deep learning technique is employed to capture fine-grained user preferences. One approach focuses on dynamically adapting user or item vectors based on influential neighbors identified for the target item. For instance, NAIS [55] employs an attention network to estimate the importance of interacted items in modeling user preferences. CMN [35] computes the target user vector by aggregating the vectors of neighboring users, with the selection and contributions of neighbors influenced by the target item. Similarly, MARS [172] adapts the target user vector based on the neighboring item vectors of the target item. Both CMN and MARS keep the target item vector fixed while adapting the user vector. Conversely, DIN [174] fixes the user preference vector and adapts the target item representation based on the user's previously purchased items.

Another major research direction leverages textual reviews to analyze user attention on various aspects of target items, incorporating these attention weights into matrix factorization for recommendation [22, 21, 26, 15]. In particular, ALFM [22] applies a topic model on reviews to detect the user attention, and A^{3}NCF [21] uses a neural attention network to learn the user attention from reviews. Following the same idea, Chin et al. [26] developed an end-to-end attentive neural network based recommendation model, which leverages reviews and ratings to learn user diverse preferences on different aspects of items. The AFM method [15] adopts a similar strategy and exploits other types of side information (*e.g.*, item category) instead of reviews. Additionally, MAML [86] introduces a multimodal attentive approach to model users' diverse preferences comprehensively.

3.2.2 Graph Convolutional Networks

Non-Euclidean structured data arise in various domains, including molecular structures in chemical analysis, sensor networks in communication systems, and social networks in the social sciences. Traditional convolutional neural networks (CNNs) are fundamentally constrained in processing such data due to its non-Euclidean nature, characterized by unstructured features and unique local neighborhoods for each node. Addressing these challenges, researchers have sought to extend the reach of deep learning into non-Euclidean domains, driven by its demonstrated success in solving complex problems in acoustics, computer vision, and natural language processing. The theoretical foundation of graph convolutional operations is rooted in the field of graph signal processing [119, 114, 115]. Bruna et al.[7] pioneered the first formulation of convolutional neural networks for graphs by leveraging spectral domain convolution defined on the graph Laplacian. Subsequent contributions by Duvenaud et al.[34], Kipf and Welling [72], and Atwood and Towsley [2] advanced the development of graph convolutional network (GCN) architectures. These models replaced traditional graph filters with adjacency matrices incorporating self-loops, employing propagation rules to compute the outputs of neural network layers during weight updates. Expanding on these advancements, Defferrard et al. [29] proposed further refinements, integrating fast localized spectral filters and efficient pooling operations to enhance computational efficiency and model performance.

In recent years, Graph Convolutional Networks (GCNs) have garnered significant attention in the realm of recommendation systems. Their rise in popularity can be attributed to their adeptness at learning node representations from non-Euclidean structures [5, 136, 135, 138, 63, 54, 87, 127, 82, 24, 141, 91]. For example, NGCF [136] propagates embeddings on the user-item interaction graph to harness the benefits of the high-order proximity. SMOG-CF [163] is proposed to directly capture the high-order connectivity between neighboring nodes at any order. Multi-GCCF [123] explicitly incorporates the user-user and item-item graphs, which is built upon the user-item bipartite graph, in the embedding learning process. Drawing inspiration from studies focused on simplifying GCNs [145], researchers also critically scrutinize the intricacies of GCN-based recommendation models. To achieve more efficient recommendations, the UltraGCN [102] model further simplifies graph convolutional networks by omitting embedding propagation layers. Additionally, it incorporates higher-order relationships through user-user graphs, leading to better performance and shorter running time compared to LightGCN. To further improve the neural graph collaborative filtering, NCL [82] explicitly incorporates the structural neighbors and semantic neighbors into contrastive pairs. Since the GCN-based recommendation methods suffer from the over-smoothing problem, IMP-GCN [87] refines user and item embeddings by conducting high-order graph convolutions within subgraphs. It can benefit from higher-order neighboring nodes by filtering out the noisy information from them. To learn more effective and robust representations, Fan et al. [37] proposed GTN, which can adaptively determine the reliability of interactions and effectively reduce unreliable interaction behaviors.

In this chapter, we argue that the existing GCN-based methods are still insufficient for obtaining valuable information from high-order neighbors since the multiple interests of users are not well considered. To tackle the problem, we proposed an unsupervised and optimizable soft node cluster method that can cluster users and items to build cluster-specific graphs. By performing high-order graph convolution on cluster-specific graphs, more valuable information could be delivered to the target node.

3.3 Preliminaries

Given a set of users $\mathcal{U}$ and a set of items $\mathcal{I}$, and interaction data (*e.g.*, ratings, click-throughs) among users and items, the goal of the recommendation system is to predict a set of items a user did not consume before and would be interested in. The interaction matrix, denoted by $\boldsymbol{R}$, is a matrix of dimensions $\mathcal{N} \times \mathcal{M}$, where $\mathcal{N}$ and $\mathcal{M}$ are the numbers of users and items, respectively. Each entry r_{ui} represents the interaction of user $u \in \mathcal{U}$ with the item $i \in \mathcal{I}$. For binary interactions, $r_{ui} = 1$ if there is an interaction, and 0 otherwise. In our method, every user and item is assigned a unique ID and is represented by an embedding vector.

User-item Interaction Graph. The User-Item Interaction Graph is a bipartite graph that provides a structured representation of the interaction data, enabling algorithms like GCNs to capture intricate patterns. Users, items, and their interactions are represented as a graph, where users and items are nodes, and interactions are edges. The aim is to learn a low-dimensional representation (or embeddings) of the nodes (*i.e.*, users and items) such that it can be used to predict future interactions. In particular, we construct the user-item interaction graph $\mathcal{G} = (\mathcal{W}, \mathcal{E})$ based on the interaction matrix $\boldsymbol{R}$, where the node set $\mathcal{W} = \mathcal{U} \cup \mathcal{I}$ consists of user nodes and item nodes, and $\mathcal{E}$ represents the edge set. In addition, we define $\mathcal{N}_u$ as the set of item nodes neighboring user node u. Similarly, $\mathcal{N}_i$ denotes the set of user nodes neighboring item node i.

3.4 ClusterGCF Model

In this section, we present **Cluster-based Graph Collaborative Filtering** (ClusterGCF) model, along with its training methodology for recommendations. As illustrated in Fig. 3.2, ClusterGCF performs the first-order graph convolution over the original user-item interaction graph [2] and the high-order graph convolution on the cluster-specific graphs to form the final representation of users and items. To construct cluster-specific graphs, we propose a soft node clustering approach to group users and items into multiple clusters. Next, based on the results of clustering, we

[2] Note that the first-order graph convolution is adopted as implemented in IMP-GCN [87].

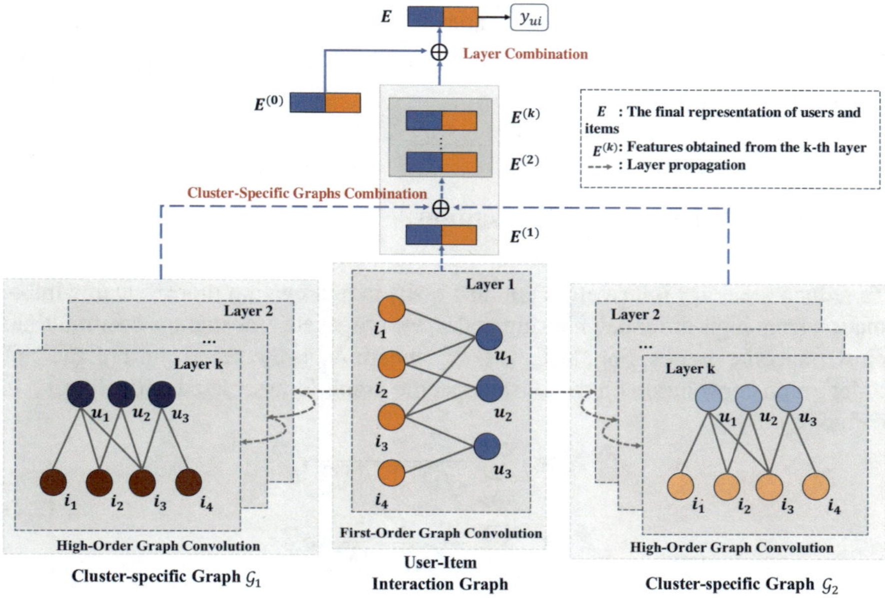

Fig. 3.2: An overview of our ClusterGCF model with two cluster-specific graphs. In ClusterGCF, first-order propagation operates on the original user-item interaction graph, and high-order propagation operates on the cluster-specific graphs $\mathcal{G}_1$ and $\mathcal{G}_2$. The cluster-specific graphs are constructed by assigning probabilities to the nodes of the original user-item interaction graph based on their association with different clusters.

assign a probability score to each node in the original user-item interaction graph for each cluster. This allows us to generate a graph that is specific to each cluster and shows the connections between users and items within that cluster. In the following, we will provide details of the ClusterGCF.

3.4.1 First-Order Graph Convolution

Since the direct interactions between users and items provide the most important and reliable information [87], all first-order neighbors are involved in the graph convolution. In ClusterGCF, the first-order graph convolution is adopted as previous studies [54, 87]:

$$
\begin{aligned}
\boldsymbol{e}_u^{(1)} &= \sum_{i \in \mathcal{N}_u} \frac{1}{\sqrt{|\mathcal{N}_u|}\sqrt{|\mathcal{N}_i|}} \boldsymbol{e}_i^{(0)}, \\
\boldsymbol{e}_i^{(1)} &= \sum_{u \in \mathcal{N}_i} \frac{1}{\sqrt{|\mathcal{N}_i|}\sqrt{|\mathcal{N}_u|}} \boldsymbol{e}_u^{(0)}.
\end{aligned}
\tag{3.1}
$$

For the target user u and item i, $\boldsymbol{e}_{\boldsymbol{u}}^{(0)}$ and $\boldsymbol{e}_{\boldsymbol{i}}^{(0)}$ are ID embeddings, while their first layer embeddings are represented as $\boldsymbol{e}_{\boldsymbol{u}}^{(1)}$ and $\boldsymbol{e}_{\boldsymbol{i}}^{(1)}$. In addition, $\frac{1}{\sqrt{|\mathcal{N}_u|}\sqrt{|\mathcal{N}_i|}}$ are symmetric normalization terms.

3.4.2 High-Order Graph Convolution

To reduce the noisy information but also assist in aggregating more valuable information from high-order neighboring nodes, we propagate embeddings over multiple cluster-specific graphs. For cluster $c \in \mathcal{N}_c$, where $\mathcal{N}_c$ is the set of clusters, the 2nd order graph convolution on the cluster-specific graph $\mathcal{G}_c$ associated with cluster c is defined as:

$$\begin{aligned} \boldsymbol{e}_{\boldsymbol{u}}^{(c,k)} &= \sum_{i \in \mathcal{N}_u} \frac{p_i^c}{\sqrt{|\mathcal{N}_u|}\sqrt{|\mathcal{N}_i|}} \boldsymbol{e}_{\boldsymbol{i}}^{(1)}, \\ \boldsymbol{e}_{\boldsymbol{i}}^{(c,k)} &= \sum_{u \in \mathcal{N}_i} \frac{p_u^c}{\sqrt{|\mathcal{N}_i|}\sqrt{|\mathcal{N}_u|}} \boldsymbol{e}_{\boldsymbol{u}}^{(1)}, \end{aligned} \tag{3.2}$$

where $\boldsymbol{e}_{\boldsymbol{u}}^{(c,k)}$ and $\boldsymbol{e}_{\boldsymbol{i}}^{(c,k)}$ denote the embedding of user u and item i after k layers graph convolution and $k = 2$. p_u^c and p_i^c represent the probabilities assigned to user u and item i for cluster c. When $k > 2$, we define the $k-$th order convolution on the cluster-specific graph $\mathcal{G}_c$ as:

$$\begin{aligned} \boldsymbol{e}_{\boldsymbol{u}}^{(c,k)} &= \sum_{i \in \mathcal{N}_u} \frac{p_i^c}{\sqrt{|\mathcal{N}_u|}\sqrt{|\mathcal{N}_i|}} \boldsymbol{e}_{\boldsymbol{i}}^{(c,k-1)}, \\ \boldsymbol{e}_{\boldsymbol{i}}^{(c,k)} &= \sum_{u \in \mathcal{N}_i} \frac{p_u^c}{\sqrt{|\mathcal{N}_i|}\sqrt{|\mathcal{N}_u|}} \boldsymbol{e}_{\boldsymbol{u}}^{(c,k-1)}. \end{aligned} \tag{3.3}$$

Cluster-Specific Graphs Combination. With the graph convolution operation, we can obtain the node representations for each layer of cluster-specific graphs. In other words, the messages are passed from the $k-$th layer neighboring nodes based on our constructed cluster-specific graphs. These messages are further combined as the $k-$th layer representations of user u and item i ($k \geq 2$):

$$\begin{aligned} \boldsymbol{e}_{\boldsymbol{u}}^{(k)} &= \sum_{c \in \mathcal{N}_c} \boldsymbol{e}_{\boldsymbol{u}}^{(c,k)}, \\ \boldsymbol{e}_{\boldsymbol{i}}^{(k)} &= \sum_{c \in \mathcal{N}_c} \boldsymbol{e}_{\boldsymbol{i}}^{(c,k)}, \end{aligned} \tag{3.4}$$

where $\boldsymbol{e}_{\boldsymbol{u}}^{(k)}$ and $\boldsymbol{e}_{\boldsymbol{i}}^{(k)}$ denote the features of user u and item i aggregated from the k-th layer neighbor nodes, respectively. Previous studies [54, 87] have demonstrated the effectiveness of aggregating information from neighboring nodes through embedding summation. Thus, we apply the same method to aggregate the information within and between cluster-specific graphs. Note that we treat the information within different cluster-specific graphs equally because assigning different probabilities to the nodes in the different clusters can capture the varying importance.

3.4.3 Layer Combination and Prediction.

Due to its effectiveness, we adopt the same layer combination and prediction methodology as LightGCN [54]. Specifically, after performing k layers of graph convolution, the final embeddings of user u and item i are obtained by combing their embeddings across all layers of the ClusterGCF:

$$\begin{aligned} \boldsymbol{e_u} &= \sum_{k=0}^{K} \alpha_k \boldsymbol{e_u^{(k)}}, \\ \boldsymbol{e_i} &= \sum_{k=0}^{K} \alpha_k \boldsymbol{e_i^{(k)}}, \end{aligned} \tag{3.5}$$

where α_k is uniformly set to $1/(K+1)$. After learning the representation for both users and items, the preference of a given user u for a target item i is estimated as follows:

$$\hat{r}_{ui} = \boldsymbol{e_u^T} \boldsymbol{e_i}. \tag{3.6}$$

3.4.4 Soft Node Clustering

We introduce a soft node clustering approach to capture the multiple interests of users and identify the common interests among them. Since users typically exhibit multiple interests in different aspects of items [86], we classify both the user and item nodes into multiple clusters with varying degrees of association instead of assigning them to a single cluster. Take a user u and an item i as an example, we start by fusing the graph structure and ID embedding:

$$\begin{aligned} \boldsymbol{F_u} &= \sigma(\boldsymbol{W_1}(\boldsymbol{e_u^{(0)}} + \boldsymbol{e_u^{(1)}}) + \boldsymbol{b_1}), \\ \boldsymbol{F_i} &= \sigma(\boldsymbol{W_1}(\boldsymbol{e_i^{(0)}} + \boldsymbol{e_i^{(1)}}) + \boldsymbol{b_1}), \end{aligned} \tag{3.7}$$

where $\boldsymbol{F_u}$ and $\boldsymbol{F_i}$ are the enhanced features of the user u and item i obtained through feature aggregation. $\boldsymbol{W_1} \in R^{d\times d}$ and $\boldsymbol{b_1} \in R^{1\times d}$ are the weight matrix and bias vector, respectively. σ denotes the activation function for which the LeakyReLU function is adopted [100]. Subsequently, we cast these enhanced features into prediction vectors using a single-layer neural network for classifying users and items into various clusters:

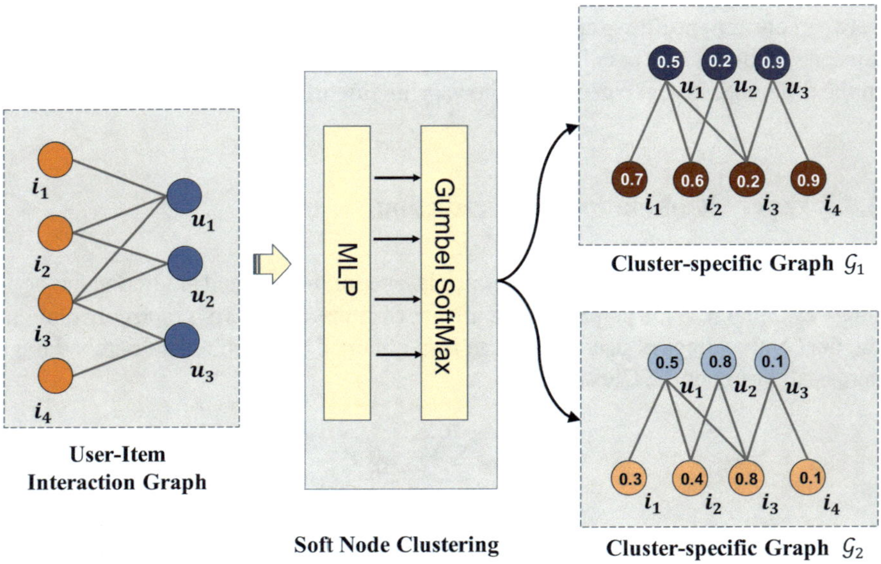

Fig. 3.3: An example of cluster-specific graphs construction. Each user and item node of the user-item interaction graph is classified into two clusters via a soft node clustering method. Specifically, each node in the original graph is allocated to the two clusters based on its soft assignment probabilities. Take user node u_2 as an example, it is assigned to cluster 1 with a weight of 0.2 and to cluster 2 with a weight of 0.8. Based on the soft assignment probabilities and the original edges from the user-item interaction graph, two cluster-specific graphs $\mathcal{G}_1$ and $\mathcal{G}_2$ are constructed.

$$\begin{aligned} \boldsymbol{H_u} &= \sigma(\boldsymbol{W_2 F_u} + \boldsymbol{b_2}), \\ \boldsymbol{H_i} &= \sigma(\boldsymbol{W_2 F_i} + \boldsymbol{b_2}), \end{aligned} \tag{3.8}$$

where H_u and H_i are the prediction vectors for the user and item, respectively. Similarly, $\boldsymbol{W_2} \in R^{d \times |\mathcal{N}_c|}$ and $\boldsymbol{b_2} \in R^{1 \times |\mathcal{N}_c|}$ are the weight matrices and bias vector. It is worth noting that the dimension of the prediction vector corresponds to the number of cluster-specific graphs, a value determined by a pre-selected hyperparameter.

Gumbel-Softmax

One of the challenges in clustering methods is the non-differentiability of discrete cluster assignments, which makes gradient-based optimization problematic. To overcome this limitation, we adopt the Gumbel-Softmax technique [64] for soft clustering in our proposed approach. Since the Gumbel-Softmax is differentiable, it can be included directly in end-to-end learning pipelines that involve neural networks. In other words, the clustering approach can be co-optimized with the recommendation model.

The Gumbel-Softmax function provides a smooth, differentiable approximation to the discrete distribution, parameterized by logits and a temperature parameter. In our approach, the last layer of the neural network generates logits for all clusters. These logits are then fed into a Gumbel-Softmax layer to produce soft cluster assignments for each user or item. For example, when given logits $\boldsymbol{H_u}$, the Gumbel-Softmax function outputs a probability distribution $\boldsymbol{P_u}$ over multiple clusters. Mathematically, it is defined as follows:

$$\boldsymbol{P_u} = \text{Softmax}\left(\frac{\boldsymbol{H_u} + g}{\tau}\right), \tag{3.9}$$

where z are the logits, g is the Gumbel noise, and τ is the temperature. The Gumbel noise can be generated through the Gumbel distribution, or more commonly, from uniform noise U using the transformation $-\log(-\log(U))$. The temperature parameter τ allows fine control over the approximation. Lower τ values approximate a hard, one-hot encoded assignment, while higher τ values produce a more uniform, exploratory distribution. The Softmax function transforms the modified logits into a probability distribution:

$$\text{Softmax}(x_i) = \frac{e^{x_i}}{\sum_j e^{x_j}}. \tag{3.10}$$

Our clustering approach can be trained end-to-end using gradient-based optimization techniques, owing to the Gumbel-Softmax function. It classifies users and items into different clusters in an unsupervised manner, without requiring ground-truth labels.

Cluster-Specific Graph Construction. After employing soft clustering techniques, specifically using a Gumbel-Softmax trick, we reconstruct the original user-item interaction graph into a series of cluster-specific graphs. Unlike conventional methods that may use thresholding to determine hard cluster assignments, our approach retains the granularity of soft assignments. Each node in the original graph is allocated to all clusters based on its soft assignment probabilities. A cluster-specific graph is defined as follows:

- **Nodes:** The user and item nodes from the original graph are weighted by their soft assignment probabilities to the given cluster.
- **Edges:** The original edges of the user-item interaction graph are retained, thus preserving the innate relationships between users and items.

The differentiation between these cluster-specific graphs lies in the weights (soft assignment probabilities) associated with each node. As shown in Fig. 3.3, each user node u_x or item node i_x is softly classified into two clusters with different probabilities. The relationship between the probabilities in the two clusters can be described as follows:

$$\begin{aligned} p_1(u_x) + p_2(u_x) &= 1, \\ p_1(i_x) + p_2(i_x) &= 1, \end{aligned} \tag{3.11}$$

where $p_1(\cdot)$ and $p_2(\cdot)$ denote the probabilities assigned to this node for these two cluster-specific graphs, respectively.

3.4.5 Model Training

Optimization

In this work, our goal is to recommend a list of items to the target users by computing their preference score. For the optimization, we adopt the pairwise learning method, which is similar to the approach employed in previous works [87]. To train our ClusterGCF model, we construct a triplet of $\{u, i^+, i^-\}$, where i^+ represents a positive item and i^- denotes a negative item. The objective function for ClusterGCF is defined as:

$$\arg\min \sum_{(\mathbf{u},\mathbf{i}^+,\mathbf{i}^-)\in\mathcal{O}} -\ln\phi(\hat{r}_{ui^+} - \hat{r}_{ui^-}) + \lambda \|\Theta\|_2^2 \tag{3.12}$$

where $\mathcal{O}$ represent the training set, defined as $\mathcal{O} = \{(u, i^+, i^-)|(u, i^+) \in \mathcal{R}^+, (u, i^-) \in \mathcal{R}^-\}$. Here, $\mathcal{R}^+$ indicates the observed interactions in the training dataset, while $\mathcal{R}^-$ denotes a set of sampled unobserved interactions. Additionally, λ represents the regularization weight, and Θ stands for the model parameters. We employ L_2 regularization to mitigate overfitting. To optimize our model and update its parameters, we utilize the mini-batch Adam technique, as described in [70].

Note that the user and item embeddings serve as inputs for both the node clustering and the recommendation system. We optimize these embeddings, which serve different purposes but are intricately linked within our optimization framework. Our approach ensures that the clustering process is continuously informed by the recommendation task, leading to clusters that are more relevant to the recommendation objective. Conversely, the embeddings benefit from the structure provided by the clusters, resulting in more robust and informative representations of users and items.

Matrix Form Propagation Rules

In order to improve the efficiency of model training, we employ a matrix form propagation rule [136] to refine the user and item representations while performing graph convolution operations. Let $\boldsymbol{E}^{(0)}$ be the representation matrix for user and item IDs, the first-layer embedding propagation of ClusterGCF is described as follows:

$$\boldsymbol{E}^{(1)} = \mathcal{L} \times \boldsymbol{E}^{(0)}, \tag{3.13}$$

where $\boldsymbol{E}^{(1)}$ is the obtained representation matrix of users and items in the first graph convolution layer and $\mathcal{L}$ is the Laplacian matrix for the user-item interaction graph.

For the higher-order graph convolution operation, the higher-order neighbor embeddings are first propagated on cluster-specific graphs. For the cluster-specific graph $\mathcal{G}_c$, $\boldsymbol{E}_c^{(k)}$ is defined as the obtained representation of users and items in the k-th cluster-specific graph convolution layer ($k \geq 2$). This process can be formulated as follows:

$$\boldsymbol{E}_c^k = \mathcal{L} \times (\boldsymbol{P_c} \odot \boldsymbol{E_c^{k-1}}), \tag{3.14}$$

where $\boldsymbol{P_c}$ denotes the probability vector of all nodes assigned to the cluster-specific graph $\mathcal{G}_c$. Then, we aggregate the embedding in the k-th layer from all cluster-specific graphs to formulate user and item representations, denoted as $\boldsymbol{E^{(k)}}$, in the k-th graph convolution layer of ClusterGCF:

$$\boldsymbol{E}^k = \sum_{c \in \mathcal{N}_c} E_c^k. \tag{3.15}$$

Finally, we combine the embeddings from all layers to obtain the final representations:

$$\boldsymbol{E} = \alpha_0 \boldsymbol{E^{(0)}} + \alpha_1 \boldsymbol{E^{(1)}} + \cdots + \alpha_K \boldsymbol{E^{(K)}}. \tag{3.16}$$

3.5 Experiments

To evaluate the effectiveness of our proposed ClusterGCF, we conducted extensive experiments on four public datasets. In particular, we primarily answer the following questions.

RQ1: Does ClusterGCF outperform state-of-the-art CF baselines in the top-n recommendation task?

RQ2: How do the number of graph convolution layers impact ClusterGCF?

RQ3: How do the number of clusters and the starting layer of Cluster-based Graph Convolution affect the performance and model Convergence of ClusterGCF?

RQ4: How do the temperature coefficient and node types affect the cluster-specific graph construction?

RQ5: How do the node representations benefit from the high-order connectivities in cluster-specific graphs?

In the following, we present the experimental setup and answer the above questions sequentially based on the experimental results.

3.5.1 Experimental setup

Data Description

To assess ClusterGCF model, extensive experiments are conducted on the following four datasets: Amazon review datasets [3], Gowalla and Yelp2018 [4].

- **Amazon Review Datasets.** This dataset stems from the extensive product review system of Amazon [86]. It includes user-product interactions, reviews,

[3] http://jmcauley.ucsd.edu/data/amazon.

[4] https://www.yelp.com/dataset.

Table 3.1: Statistics of the Experimental Datasets.

Dataset	#user	#item	#interactions	sparsity
Office	4,874	2,406	52,957	99.55%
Kindle Store	14,356	15,885	367,477	99.83%
Gowalla	29,858	40,981	1,027,370	99.92%
Yelp2018	31,668	38,048	1,561,406	99.87%

and metadata. The interaction information is utilized in our experiments. Two product categories from the 5-core version of this dataset are selected for evaluation: Office and Kindle Store. This indicates that both users and items have a minimum of 5 interactions.

- **Gowalla.** This dataset is collected from Gowalla [5], where users share their locations by checking-in [136]. It typically contains several millions of check-ins spread across users and locations. Each check-in also associates a user with a particular location, which is used to model user preferences in our experiments.
- **Yelp2018.** This dataset is a subset of the Yelp challenge of 2018. It contains a variety of information mainly focused on businesses, reviews, and users. Businesses, such as restaurants and services, are treated as the items in our experiments [136].

To ensure the data quality, we filtered users and items with limited interactions following the general setting in the recommendation [56, 86]. Specifically, for the Kindle Store, Gowalla, and Yelp2018 datasets, we implemented a 10-core setting, retaining only users and items with a minimum of 10 interactions. The details and statistics of these four datasets are shown in Table 3.1. From the table, the datasets vary in size, which allows us to analyze our method and its competitors' performance under different scenarios. For each dataset, we randomly split each user's interactions into training and testing sets at an 80:20 ratio. We select 10% of interactions in the training set randomly to create the validation set for tuning hyper-parameters. The pairwise learning strategy is adopted for parameter optimization. Specifically, the observed user-item interactions were considered positive instances in the training set. For each positive instance, we paired it with a random sampling negative instance that the user has not interacted with yet. For a fair comparison, we consistently adopt the negative sampling strategy across our methods and competitors.

Evaluation metrics

In this work, we adopted three metrics for the evaluation: **Recall@K** [87], **HR@K (Hit Ratio)** [56], and **NDCG@K (Normalized Discounted Cumulative Gain)** [52]. Below, we provide detailed explanations of *Hit Ratio*, as *Recall* and *NDCG* have already been introduced in Chapter 2.

[5] https://www.gowalla.com.

Hit Ratio measures whether at least one of the recommended items is relevant to the user. For a given user u, **HR** is defined as:

$$\mathrm{HR} = \begin{cases} 1 & \text{if } \mathrm{Rel}(u) \cap \mathrm{Rec}(u, n) \neq \emptyset, \\ 0 & \text{otherwise.} \end{cases} \tag{3.17}$$

This metric is often aggregated across all users as the proportion of users for whom at least one relevant item appears in the top-n recommendations.

For each metric, we assess the performance of our method and all competitors based on the top 20 items in the ranking list, with $K = 20$. Notice that the reported results are the average values for all users in the testing set.

Compared Baselines

To verify the effectiveness of ClusterGCF, we compared it with several competitive methods, mainly considering representative GCN-based models like GCMC, NGCF, LightGCN and UltraGCN, as well as those learning with different advanced techniques, such as disentangled learning (*i.e.*, DGCF), contrastive learning (*i.e.*, NCL), or those also considering noise information like GTN and IMP-GCN.

- **NeuMF [56]**: This stat-of-the-art neural collaborative filtering method captures the non-linear interactions between user and item features by introducing multiple hidden layers over the combined user and item embeddings.
- **GCMC [5]**: This method uses a single graph convolutional layer based on the GCN technique [71] to take advantage of direct connections between users and items.
- **NGCF [136]**: This is the very first GCN-based recommendation model that incorporates high-order connectivity of user-item interactions. It encodes the collaborative signal from high-order neighbors by performing embedding propagation on the user-item bipartite graph.
- **LightGCN [54]**: This model can be considered a simplified version of NGCF [136], where certain components such as the feature transformation and the non-linear activation module are removed. The model structure's simplicity can help avoid overfitting, especially when data is not massive.
- **UltraGCN [102]**: This model simplifies LightGCN [54] by eliminating infinite layers of message passing, leading to more efficient recommendations. Specifically, it approximates multiple graph convolution layers and incorporates different node relationships flexibly through a constraint loss.
- **DGCF [138]**: This GCN-based method refines the interaction graphs considering the various user intents. Meanwhile, it encourages the independence of representations with different intents by adopting the disentangled representation technique.
- **IMP-GCN [87]**: This method conducts high-order graph convolution on subgraphs, constructed based on users with shared interests and associated items.

By differentiating high-order neighbors, it could alleviate the over-smoothing problem in GCN-based methods.
- **NCL [82]**: This is a model-agnostic contrastive learning framework. It seeks to capture the correlation between a node and its prototype to improve the neural graph collaborative filtering.
- **GTN [37]**: This GNN-based model captures the adaptive reliability of interactions between users and items, drawing inspiration from the concept of trend filtering and graph trend filtering.

Parameter settings

We used Tensorflow [6] to implement ClusterGCF. For all models, the embedding size was fixed at 64 and initialized using Xavier [149]. For optimization, we used Adam [70] with a default learning rate of 0.001. The mini-batch size is 1024 on Office and Kindle Store. In contrast, it is increased to 2048 on Gowalla and Yelp208 to accelerate the model training. The L_2 regularization coefficient λ is searched in $\{1e^{-6}, 1e^{-5}, \cdots, 1e^{-1}\}$. The embedding propagation layer is searched in the range of $\{1, 2, \cdots, 8\}$. The GCMC method uses a single convolutional layer as its original setting. For ClusterGCF and IMP-GCN, the cluster number is searched in the range of $\{2, 3, 4\}$. Besides, the temperature coefficient is tuned amongst $\{1e^{-2}, 1e^{-1}, \cdots, 1e^{2}\}$. The early stopping and validation strategies used are identical to those employed in IMP-GCN, ensuring that we carefully tuned the key parameters.

3.5.2 Performance Comparison (RQ1)

Overall comparison

In this section, we compared the performance of our ClusterGCF with all the adopted competitors. Table 3.2 shows the experimental results in terms of Recall@20, HR@20, and NDCG@20 for ClusterGCF and all compared methods over four different datasets. The top two performing results are bolded for clarity. From these results, several key observations arise:

NeuMF is a DNN-based model and does not perform as effectively as the GNN-based models tested in our experiments. This shortfall can be attributed to its lack of explicit consideration for both first- and higher-order connectivity between users and items in representation learning. In contrast, graph-based approaches like GCMC perform better than NeuMF, demonstrating the benefits of leveraging graph structure. However, GCMC underperforms NGCF, because it is limited to utilizing only first-order neighbor nodes. As the very first GCN-based collaborative filtering model, NGCF brought into play the high-order connectivity of user-item interactions through GCN techniques. This innovation in embedding high-order connectivities offers a

[6] https://www.tensorflow.org.

Table 3.2: Performance comparison between our ClusterGCF model and the baselines on four datasets. Please note that the values are presented as percentages, with the '%' symbol omitted.

Datasets	Office			Kindle Store			Gowalla			Yelp2018		
Metrics	Recall	HR	NDCG	Recall	HR	NDCG	Recall	HR	NDCG	Recall	HR	NDCG
NeuMF	6.92	21.04	6.29	6.83	20.19	4.98	12.95	45.65	11.18	4.70	31.12	3.72
GCMC	7.15	21.76	6.85	7.73	20.88	5.22	14.07	48.86	11.78	4.82	31.89	3.87
NGCF	12.57	32.44	7.42	9.41	25.39	5.68	15.70	52.78	13.27	5.81	36.64	4.75
LightGCN	13.19	33.20	7.83	9.94	26.53	6.08	18.09	58.83	15.42	6.33	38.52	5.12
UltraGCN	13.52	33.87	8.13	10.70	28.49	6.63	18.60	59.60	15.72	6.70	40.51	5.47
DGCF	13.27	33.23	7.89	10.05	27.05	6.15	18.24	58.83	15.51	6.37	38.83	5.08
IMP-GCN	13.48	33.84	8.11	10.77	28.61	6.75	18.62	59.77	15.76	6.66	40.43	5.45
NCL	13.40	33.35	8.08	10.65	28.22	6.57	18.57	59.43	15.66	6.63	40.12	5.38
GTN	13.54	34.14	8.16	10.78	28.76	6.78	18.65	59.97	15.83	6.75	40.75	5.54
ClusterGCF	**13.97***	**35.02***	**8.46***	**11.16***	**29.57***	**7.11***	**19.07***	**61.31***	**16.21***	**7.06***	**42.36***	**5.83***
Improv.	3.18%	2.58%	3.68%	3.53%	2.82%	4.87%	2.25%	2.23%	2.4%	4.59%	3.95%	5.23%

The symbol * denotes that the improvement is significant with $p - value < 0.05$ based on a two-tailed paired t-test.

performance boost. LightGCN consistently outperforms both GCMC and NGCF. It simplifies the NGCF by eliminating redundant operations and reducing computational complexity. This showcases the strength of GCN and emphasizes the significance of integrating high-order information into representation learning. UltraGCN further simplifies the GCNs. It skips the embedding propagation layers, leading to highly efficient recommendation. As a result, UltraGCN surpasses LightGcn across all datasets.

DGCF yields better performance than LightGCN because it uses a disentangled representation to learn independent and robust user and item embeddings by taking into account the diverse intents of users. Both IMP-GCN, NCL and GTN consistently outperform all the above GCN-based methods over all the datasets. This is largely due to their ability to effectively filter out noisy information from neighboring nodes. The enhanced performance of IMP-GCN over LightGCN and DGCF emphasizes the significance of differentiating nodes among high-order neighbors during the graph convolution operation. On the other hand, GTN achieves promising results by capturing the adaptive reliability of the interaction between users and items, which helps in reducing the negative impact of unreliable information.

Our ClusterGCF method consistently surpasses all baseline models across all datasets in terms of all metrics. It effectively avoids the noise and captures more valuable information from high-order neighbor nodes, alleviating the over-smoothing problem in the GCN-based methods. Moreover, the improvement over IMP-GCN demonstrates the effectiveness of our node clustering method. Its superior performance can be attributed to the following reasons: (1) ClusterGCF considers multiple interests of each user while performing high-order graph convolutions. (2) The pro-

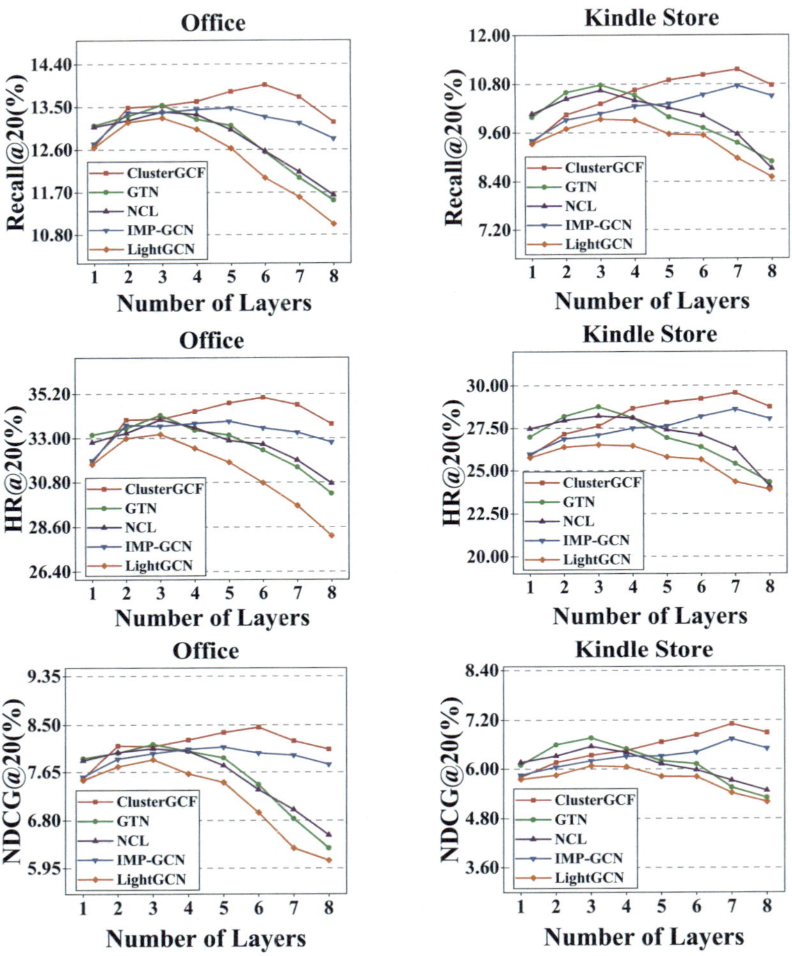

Fig. 3.4: Performance comparison between ClusterGCF and Competitors at different layers on Office and Kindle Store.

posed soft node clustering approach effectively captures the multiple interests of users, identifies the common interests among them, and can be optimized with the recommendation model in an end-to-end manner.

3.5.3 Effect of Graph Convolution Layers (RQ2)

In this section, we evaluated the performance of our ClusterGCF model by varying the number of layers in the graph convolution process. Our primary goal is to examine whether the ClusterGCF model can mitigate the over-smoothing problem

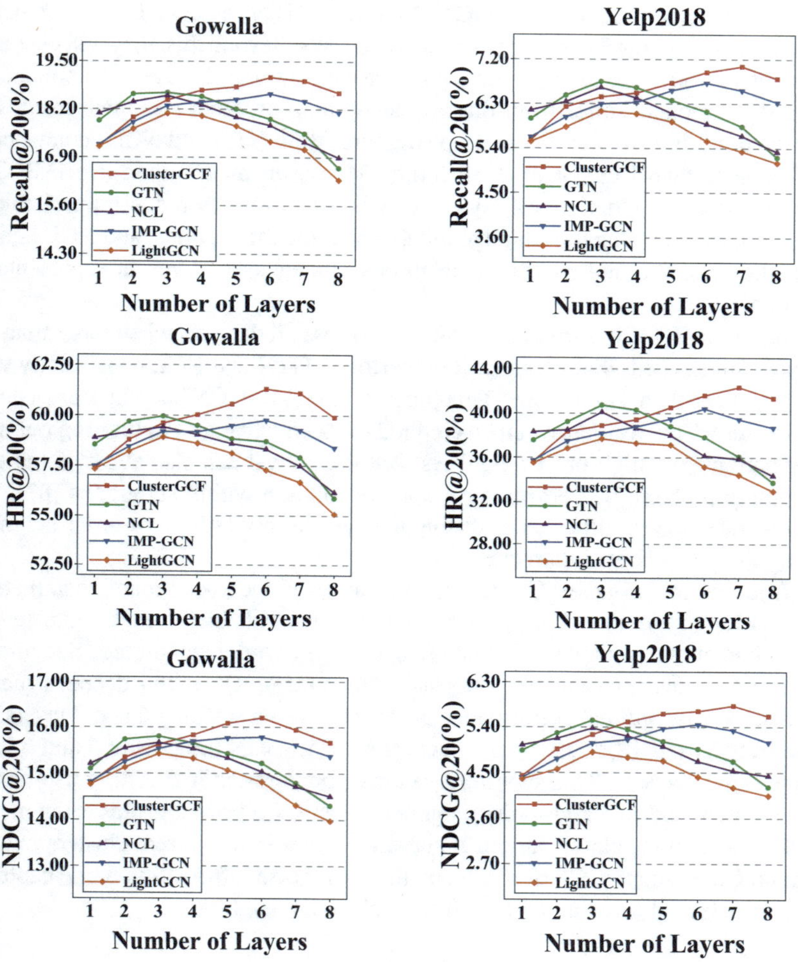

Fig. 3.5: Performance comparison between ClusterGCF and Competitors at different layers on Gowalla and Yelp2018.

often encountered in GCN-based recommendation models. To investigate the effectiveness of ClusterGCF, we iteratively increased the number of convolution layers from 1 to 8. We then compared its performance against four GCN-based methods: LightGCN, IMP-GCN, NCL, and GTN. The comparative outcomes are depicted in Fig. 3.4 and 3.5. In particular, both ClusterGCF and IMP-GCN attained their optimal performance while performing first-order graph convolution across the entire graph. IMP-GCN performs high-order graph convolution within the subgraphs (see Section 3.5.4) while ClusterGCF operates within the cluster-specific graphs. As a result, both two models adopt the same message-passing strategy as LightGCN in the first-order convolution layer. From the results, we had some interesting observations.

First, all three models (LightGCN, NCL, and GTN) achieve their peak performance when stacking 3 or 4 layers across all datasets. Beyond this, they will encounter dramatic performance degradation with increasing layers. These results indicate that these three models suffer from the over-smoothing problem and inadvertently introduce noise information in the deep structure. Both GTN and NCL achieve better performance than LightGCN at all layers. Moreover, they outperform IMP-GCN when stacking less than 3 or 4 layers. This is because GTN reduces unreliable information by capturing the adaptive reliability of the interactions, and NCL benefits from the semantic and structural neighbors among users/items in representation learning.

Second, IMP-GCN surpasses LightGCN across all datasets when more than 2 or 3 layers are stacked. Significantly, it outperforms GTN and NCL consistently when stacking more than 4 or 5 layers. This suggests that IMP-GCN benefits from a deeper architecture. This resilience can be credited to its effectiveness in filtering out noisy information from high-order neighbors. Additionally, it can also alleviate the over-smoothing problem by performing graph convolution within subgraphs [87]. This demonstrates the significance of distinguishing user nodes in high-order neighbors during the graph convolution operation.

Moreover, the proposed ClusterGCF surpasses all the above-mentioned methods after stacking more than 4 layers across all datasets. This indicates that ClusterGCF can obtain more valuable information from higher-order neighbors. Specifically, ClusterGCF can achieve better performance than IMP-GCN with a deeper structure. For instance, its optimal performance is observed when stacking 6 and 7 layers over Office and Yelp2018, respectively. In contrast, IMP-GCN peaks with 5 and 6 layers on the same datasets. This superiority stems from ClusterGCF distinguishing both user and item nodes in high-order neighbors. Besides, a learnable clustering method is adopted for node classification. It classifies nodes into different clusters in a soft manner. Consequently, ClusterGCF can filter out noise information while capturing more valuable information passing from high-order neighbors.

3.5.4 Effect of Cluster-specific Graphs (RQ3)

In this section, we study the effects of cluster-specific graphs (numbers and the starting layer of graph convolution on the cluster-specific graphs) on the performance of our ClusterGCF. Besides, we also analyze the model Convergence of ClusterGCF to verify the effectiveness of employing cluster-specific graphs.

Cluster-specific Graph Numbers

To verify the effectiveness of our node clustering approach, we evaluated ClusterGCF and IMP-GCN across various numbers of clusters, specifically $\{2, 3, 4\}$, on all datasets. The experimental results are illustrated in Fig. 3.6 and 3.7.

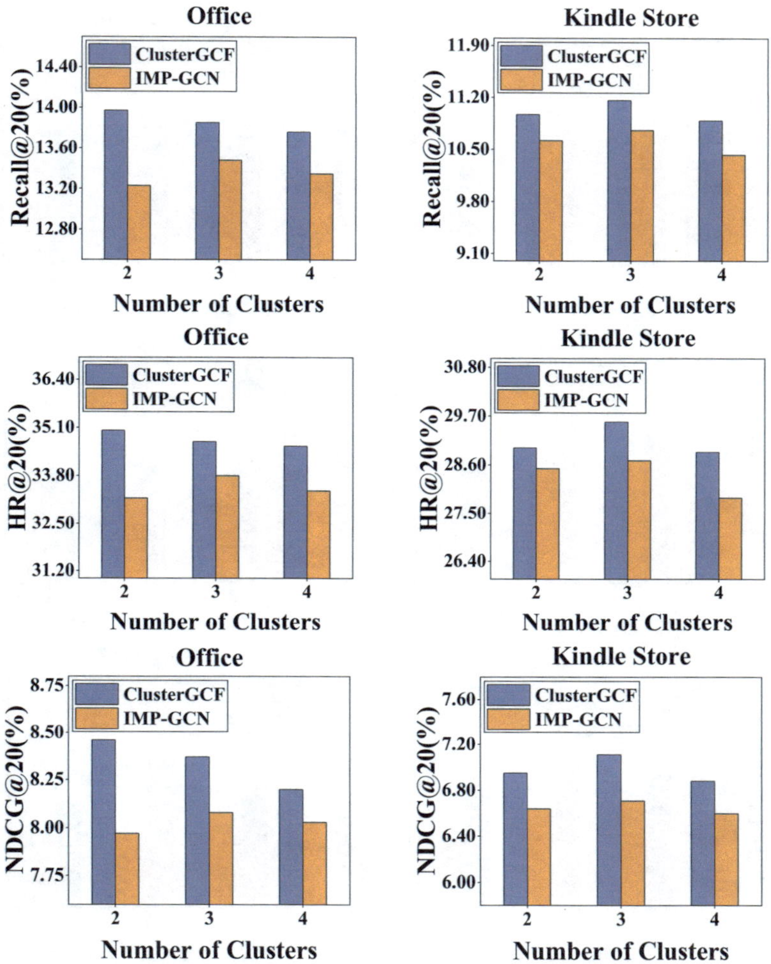

Fig. 3.6: Performance Comparison between ClusterGCF and IMP-GCN with different cluster numbers on Office and Kindle Store.

From the results, we can see that ClusterGCF outperforms IMP-GCN across all cluster number settings. This can be attributed to several reasons: (1) **Item Node Clustering**. ClusterGCF offers an advantage by distinguishing item nodes among neighbors, which effectively reduces noise while improving valuable information from high-order neighbors. Specifically, IMP-GCN only uses users of common interest and their associated items to build subgraphs. In contrast, ClusterGCF builds cluster-specific graphs considering both the users with common interests and the items that appeal to them. (2) **Soft Clustering**. In ClusterGCF, the nodes (user nodes or item nodes) are assigned weights that denote the probabilities of their association with a particular cluster. This approach allows a node presence in multiple clusters.

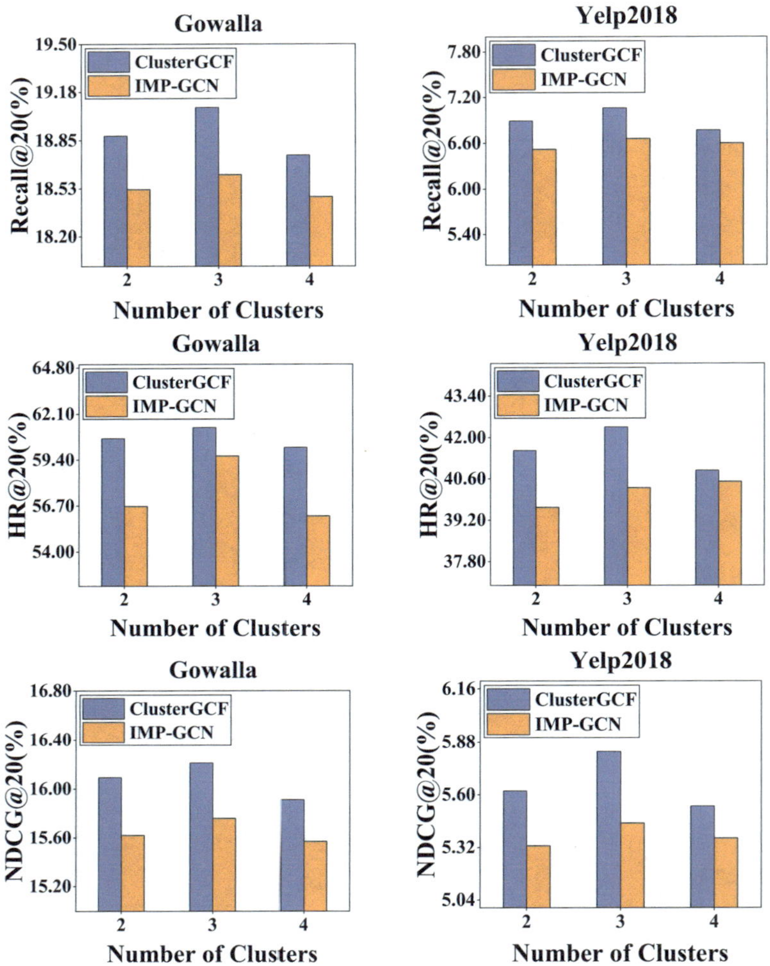

Fig. 3.7: Performance Comparison between ClusterGCF and IMP-GCN with different cluster numbers on all Gowalla and Yelp2018.

Thus, several cluster-specific graph might share some important nodes, enhancing the quality of information in the embedding propagation operation. (3) **Optimizable**. In comparison to IMP-GCN, the adopted node clustering method is optimized. The node clustering approach is co-trained with the recommendation model.

Another interesting finding is that ClusterGCF achieves its peak performance with 2 cluster-specific graphs on Office. This is because there is a trade-off in selecting the number of cluster-specific graphs [87]. Having more cluster-specific graphs allows ClusterGCF to distinguish users and items at a more finer level, which enables it to distill information more effectively from high-order neighbors. However, it also reduces the valuable information from the neighbors in short-distance in embed-

Table 3.3: Performance comparison between ClusterGCF and its variants on four datasets. Please note that the values are presented as percentages, with the '%' symbol omitted.

Datasets	Office			Kindle Store			Gowalla			Yelp2018		
Metrics	Recall	HR	NDCG	Recall	HR	NDCG	Recall	HR	NDCG	Recall	HR	NDCG
ClusterGCF$_F$	13.78	34.65	8.25	10.94	28.91	6.8	18.74	60.44	15.92	6.74	40.72	5.55
ClusterGCF$_T$	13.86	34.78	8.40	11.05	29.33	6.95	18.88	60.71	16.09	6.86	41.18	5.61
ClusterGCF	**13.97**	**35.02**	**8.46**	**11.16**	**29.57**	**7.11**	**19.07**	**61.31**	**16.21**	**7.06**	**42.36**	**5.83**

ding learning. Especially, IMP-GCN consistently underperforms ClusterGCF across all cluster-specific graph numbers, further highlighting the effectiveness of passing messages with considering users' multiple interests.

The Starting Layer of Cluster-based Graph Convolution

To investigate the effect of the starting layer of cluster-based graph convolution, we devised the following two variants:

- **ClusterGCF$_F$**: This variant performs both the first- and high-order graph convolution on the cluster-specific graphs.
- **ClusterGCF$_T$**: In this variant, both the first- and second-order embedding propagation are conducted on the original user-item interaction graph.

Table 3.3 presents the results of two variants as well as ClusterGCF. The experimental results show that ClusterGCF$_T$ performs better performance than ClusterGCF$_F$. This indicates that the neighbors of short distance play a crucial role in the representation learning of the target node. In other words, the poor performance of ClusterGCF$_F$ is because distinguishing the first-order neighbors at a finer level may diminish the valuable information they provide. The result of ClusterGCF$_T$ also indicates the effectiveness of differentiating the high-order neighbor nodes. Across all datasets, ClusterGCF consistently outperforms both ClusterGCF$_F$ and ClusterGCF$_T$. This means that our ClusterGCF can achieve its peak performance on these three datasets when we perform second- and higher-order graph convolution on the cluster-specific graphs, as demonstrated by IMP-GCN [87]. More specifically, the information from the second-order neighbors may introduce the noise information, resulting in sub-optimal performance.

Model Convergence

We analyze the convergence of ClusterGCF and IMP-GCN based on their performance in terms of Recall over four datasets.

As illustrated in Fig. 3.8, ClusterGCF exhibits faster convergences than IMP-GCN. For example, ClusterGCF reaches peak performance at 320 epochs, while IMP-GCN requires 360 epochs on the Kindle Store dataset. This can be attributed to

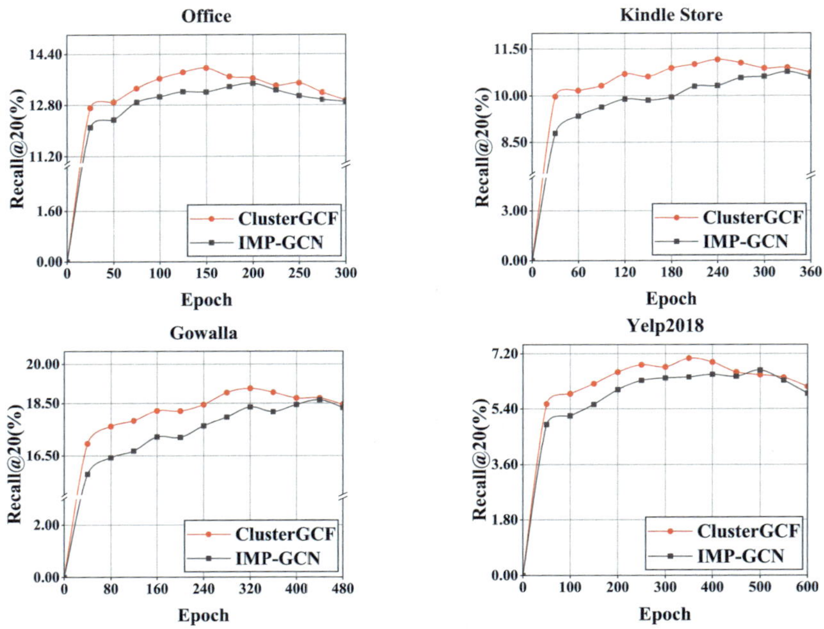

Fig. 3.8: Performance evaluation for each epoch of ClusterGCF and IMP-GCN.

the soft clustering method employed in ClusterGCF. Soft clustering provides more flexibility for the model. When the model tries to find the optimal classification or clusters, it does not need to forcefully assign a node to a specific cluster but can instead assign a probability to it. In other words, the model can adjust its decision boundaries more smoothly, accelerating the learning process. Due to the smooth nature of soft clustering, the model might encounter fewer local optima and traps. Hence, the optimization algorithm could more easily find a good solution. In summary, soft clustering offers smoother and more informative feedback, aiding the model in learning and converging more rapidly.

3.5.5 Effect of Node Clustering (RQ4)

In this section, we examine the impact of the temperature coefficient on node clustering. Additionally, we investigate the influence of node type in clustering and the effect of the clustering method on cluster-specific graph construction.

Temperature Coefficient

In our ClusterGCF model, the Gumbel-Softmax technique plays a pivotal role in facilitating soft clustering for user and item nodes. A key component in this method is the

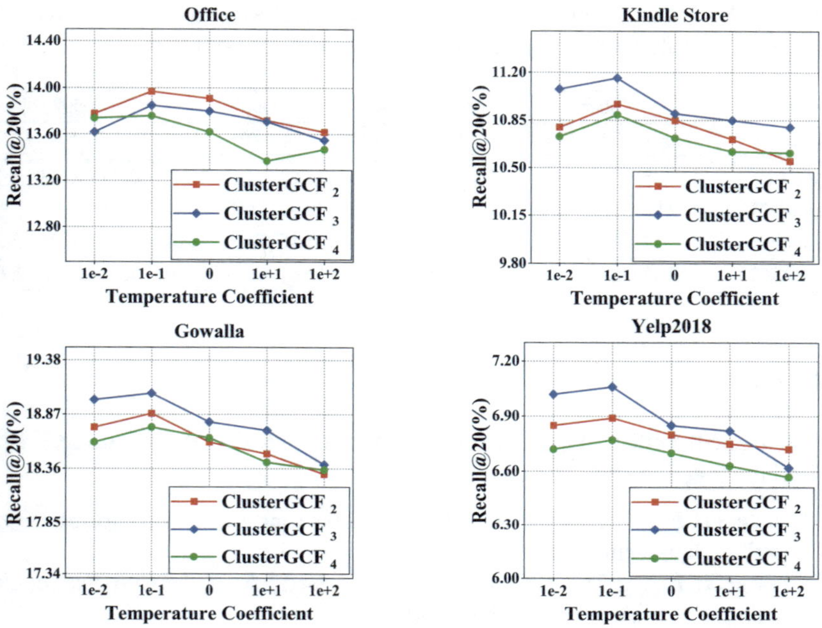

Fig. 3.9: Effect of Temperature Coefficient on the Performance of ClusterGCF Models. ClusterGCF_2, ClusterGCF_3 and ClusterGCF_4 represent ClusterGCF with 2, 3, and 4 cluster-specific graphs,respectively.

temperature coefficient, which controls the smoothness of the output. Lower values make the output closer to one-hot encoded vectors (hard classification), while higher values make the output more uniform (soft classification). To investigate the impact of the temperature coefficient on ClusterGCF, we performed experiments by varying the value of the temperature coefficient in the range of $\{1e^{-2}, 1e^{-1}, \cdots, 1e^{2}\}$. Then, we evaluated the model's performance with different cluster-specific graph numbers on different datasets in terms of *Recall* [7].

Fig. 3.9 illustrates the performance of ClusterGCF on all datasets. From the results, with increasing of the temperature coefficient, there was a noticeable improvement in performance. This is because more valuable information can be passed from high-order neighbors. The clusterGCF reached its peak performance when the temperature coefficients were set to $1e^{-1}$. After that, further amplification in the coefficient beyond this threshold led to a performance drop-off. This is because higher values make the clustering results uniform. Thus, ClusterGCF cannot differentiate nodes among high-order neighbors, resulting in suboptimal performance.

[7] The performance of ClusterGCF with respect to *HR* and *NDCG* show a similar trend as *Recall*.

Table 3.4: Performance of IMP-GCN, ClusterGCF and its variants over four datasets. Notice that the values are reported by percentage with '%' omitted.

Datasets	Office			Kindle Store			Gowalla			Yelp2018		
Metrics	Recall	HR	NDCG	Recall	HR	NDCG	Recall	HR	NDCG	Recall	HR	NDCG
IMP-GCN	13.48	33.84	8.11	10.77	28.61	6.75	18.62	59.77	15.76	6.66	40.43	5.45
ClusterGCF$_{user}$	13.88	34.95	8.38	11.03	29.12	7.03	18.97	61.07	16.09	6.94	41.85	5.67
ClusterGCF$_{item}$	13.72	34.51	8.22	10.84	28.77	6.82	18.85	60.60	16.01	6.82	40.04	5.50
ClusterGCF	**13.97**	**35.02**	**8.46**	**11.16**	**29.57**	**7.11**	**19.07**	**61.31**	**16.21**	**7.06**	**42.36**	**5.83**

Node Types

To investigate the effect of node type involved in node clustering for ClusterGCF, we devised the following two variants:

- **ClusterGCF**$_{user}$: This variant solely classifies user nodes into various clusters based on their shared interest.
- **ClusterGCF**$_{item}$: This variant solely categorized item nodes into distinct clusters based on their common characteristics.

In this section, we investigate different node types to validate the effectiveness of node clustering within ClusterGCF model. Specifically, the ClusterGCF$_{user}$ focuses on clustering user nodes, while ClusterGCF$_{item}$ is dedicated to item nodes. We present a comparative analysis of our ClusterGCF, its two specific variants, and IMP-GCN. The experimental results are detailed in Table 3.4. These results offer several intriguing insights.

First, ClusterGCF$_{user}$ achieved better performance than IMP-GCN. This is because ClusterGCF$_{user}$ benefits from the adopted clustering approach which is optimizable and it clusters user nodes in a soft manner. As a result, the target node can obtain more valuable information from high-order neighbors in representation learning. Second, ClusterGCF$_{item}$ also performs better than IMP-GCN, indicating the effectiveness of item node clustering within ClusterGCF. Lastly, ClusterGCF consistently outperforms the ClusterGCF$_{user}$ and ClusterGCF$_{item}$ across all datasets. This demonstrates the robustness and effectiveness of the node clustering approach in ClusterGCF.

3.5.6 Visualization (RQ5)

To further analyze the effect of ClusterGCF on representation learning within different graph convolution layers, we visualized the embeddings propagated from neighbor nodes of varying orders. Specifically, we first classify the user nodes into three clusters using the soft clustering method over the Kindle Store. Then, we visualized the embedding propagated from high-order neighbor nodes, using the

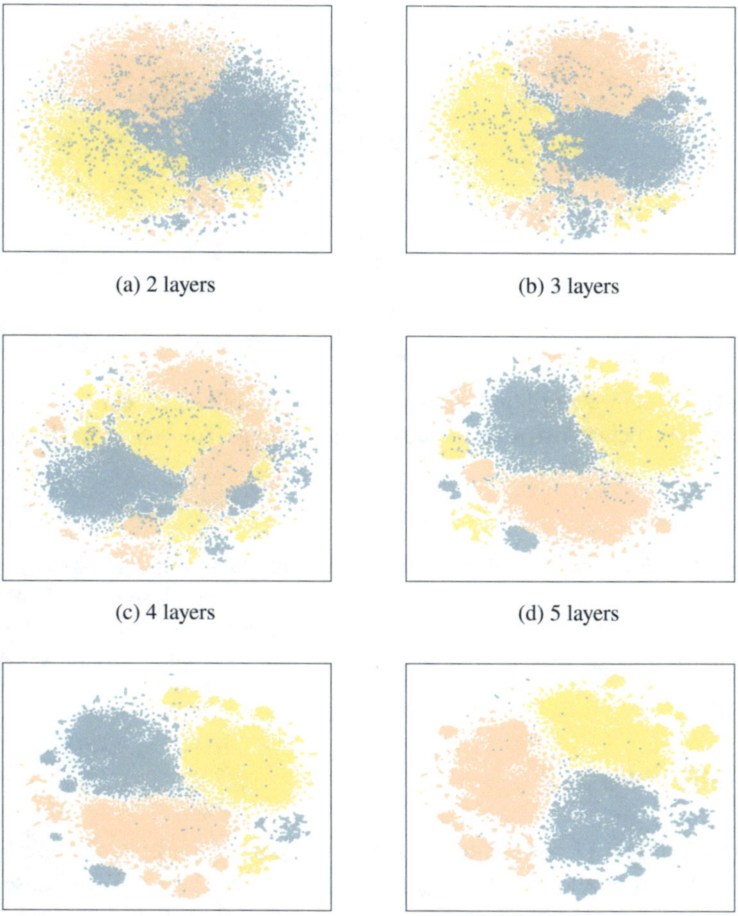

(a) 2 layers (b) 3 layers

(c) 4 layers (d) 5 layers

(e) 6 layers (f) 7 layers

Fig. 3.10: Visualization of propagated embeddings within different graph convolution layers for the target user nodes, where each color corresponds to a cluster.

t-SNE [101] in Fig. 3.10. The figures show the embedding of neighbors of different orders, ranging from 2 to 7. Each dot in the figures denoted the embedding propagated from a user node, while dots of the same color across all figures denote the same cluster of users. Fig. 3.10 shows the following interesting observations.

As graph convolution layers increase, the embedding vectors derived from high-order neighbors exhibit growing independence across different clusters. For instance, at the 2-order level, while embeddings within each cluster manifest specific aggregation tendencies, they are still comparatively scattered. With the addition of more layers, there's an evident shift in the node embeddings within each cluster. By the

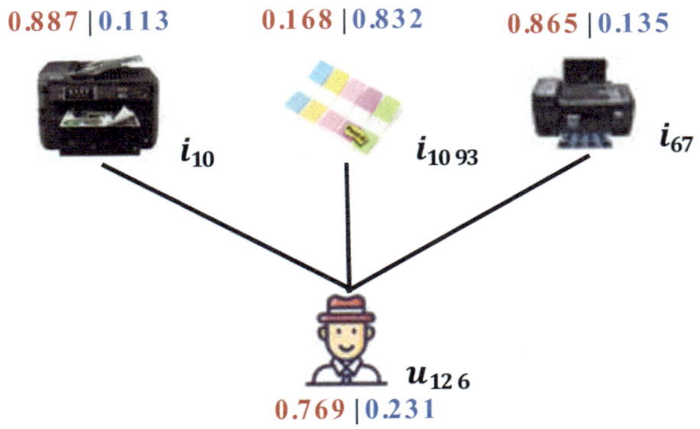

Fig. 3.11: An example of high-order neighbor nodes in two cluster-specific graphs. The weights associated with each node are marked in red (cluster 1) and blue (cluster 2).

7th layer, the dispersion notably diminishes, resulting in embeddings that display heightened independence. The representation learning in ClusterGCF is enhanced by these independent embeddings from high-order neighbors, as such independent representations have proven to be more robust against complex variations [99].

3.5.7 Case Study

As shown in Fig. 3.11, we conducted a case study to demonstrate the effectiveness of our proposed method in filtering out noisy neighborhoods. We choose a user node u_{126} at the second graph layer and the interacted three item nodes i_{10}, i_{67} and i_{1093} on Office. The weights associated with each node within two cluster-specific graphs are marked in red (cluster 1) and blue (cluster 2). Although the user u_{126} interacted with three items, two of them (i_{10} and i_{67}) are assigned greater weights in cluster 1, while the other item (i_{1093}) is assigned a greater weight in cluster 2. Consequently, during the embedding propagation process, the influence of this node's information is reduced in both cluster-specific graphs, demonstrating that our method can effectively filter out noisy information.

3.6 Conclusion

In this chapter, we argued that existing GCN-based methods are inadequate for obtaining reliable and adequate information from high-order neighboring nodes since the multiple interests of users are not properly considered. To alleviate this issue, we proposed the ClusterGCF model that performs high-order graph convolution inside cluster-specific graphs constructed by exploiting common interests among users and their multiple interests. To group both user and item nodes into multiple clusters, we introduced a soft node clustering method, which is unsupervised and optimizable, to estimate the probability assigned to the node for each cluster. The cluster-specific graphs are constructed relying on the topology of the user-item interaction graph and the clustering results. Since the embeddings of nodes are propagated within cluster-specific graphs, ClusterGCF can filter out the negative information while capturing more valuable information from high-order neighboring nodes. Besides, the proposed clusterGCF maintains the uniqueness of node embeddings while stacking more graph convolution layers because it can benefit from higher-order neighboring nodes in cluster-specific graphs. Experiments on four datasets demonstrate that ClusterGCF can achieve state-of-the-art performance. The superior performance of ClusterGCF indicates the importance of distinguishing high-order neighbors in enhancing representation learning and demonstrates the effectiveness of the proposed soft node clustering approach.

This method attempts to exploit the common interest between users and the multiple interests of a user by grouping the users and items in the high-order graph convolution operation, limited to the unsupervised clustering approach and recommendation scenario. Specifically, there are many other clustering approaches, including using supervised information or additional knowledge. Furthermore, the cluster-based graph convolution operation can be widely applied in various tasks, such as social network analysis and information retrieval. We expect the potential of the cluster-based GCNs to be further explored towards other tasks.

Chapter 4
Semantic Aspect-aware Graph Convolutional Network

4.1 Introduction

As discussed in Chapter 2 and 3, users often exhibit common or multiple interests, which adds complexity to understanding and modeling their preferences. This diversity not only highlights the importance of interpretability in recommendation systems but also underscores the challenges in achieving it. User reviews, rich in sentiment and contextual details, offer a valuable source of information for uncovering this complexity. In this chapter, we aim to harness the expressive power of reviews to explicitly extract nuanced insights into users' fine-grained preferences. By integrating this information into recommendation models, we seek to enhance their interpretability, enabling more transparent and user-centric recommendations that align with users' explicit and implicit preferences.

To better understand user preferences, researchers have paid significant attention to exploiting fine-grained behaviors between users and items on different aspects or intents [22, 21, 26, 86, 13]. Existing methods can be categorized into two main approaches. First, employing topic models [6], the statistical model designed to identify the abstract "topics" based on the frequency and co-occurrence of words in textual content, to extract aspects from these reviews. These extracted aspects are then incorporated into user preference modeling [22, 21]. In particular, ALFM [22] applies a topic model on reviews to extract the topics the user pays attention to, while A^{3}NCF [21] learns the user attention from reviews via a neural attention network in an end-to-end manner. While these models can identify topics or aspects in the review, understanding their semantic meaning is not straightforward. In addition, the noise in the review has a negative impact on the aspect extraction. For example, words that have no contribution to user preference, like "the", "last", and "year" in reviews, are extracted by the topic models and leveraged into the training process of the recommendation model. Second, disentangling the underlying aspect/factor behind interaction data [139, 138, 83]. Disentangled representations are becoming increasingly important for learning robust and independent representations from user-item interaction data in recommendation [139, 138]. For example, DisenHAN [139] learns

F. Liu, L. Nie, *Advancing Recommender Systems with Graph Convolutional Networks*,
https://doi.org/10.1007/978-3-031-85093-6_4

disentangled user/item representations from different aspects in a heterogeneous information network, while DGCF [138] models a distribution over intents for each user-item interaction to study the diversity of user intents on adopting the items. Despite their ability to accurately model user preference and capture the underlying factors, such models extract the aspects relying on user-item interaction data, which suffer from data sparsity and bias issues in recommendations. In addition, the implicit aspects and intents behind the recommendations are often elusive and not readily interpretable.

Semantic aspects often denote specific features or attributes of a product or service in which users might be interested in the context of recommendation systems and sentiment analysis. Extracting semantic aspects from reviews could enhance our comprehension of the complex relations between each user and item at a finer level. Moreover, semantic aspects, like meaningful attributes or concepts of an item, provide a deeper understanding of its inherent features, thereby enhancing the interpretability of recommendations. As shown in Fig. 4.1, the review might mention various semantic aspects such as *functionality*, *durability*, and *ease of use*. Each of these aspects carries a specific meaning related to user experience. By analyzing and summarizing the raw review into structured semantic aspects, we can better understand the underlying rationales behind user behaviors. Focusing on reviews that highlight specific item aspects enables us to discern semantic aspect-aware interactions. For instance, a review might emphasize aspects like *functionality* and *durability* but omit *ease of use*. Such patterns indicate distinct interactions between the user and item concerning *functionality* and *durability*, but not for *ease of use*. Utilizing these semantic-aware insights leads to richer and more accurate representations of both users and items.

Large Language Models (LLMs) have emerged as powerful tools in natural language processing and have recently gained significant attention in various fields [28, 120]. Their contextual understanding allows for distilling various semantic aspects based on reviews without being affected by noise present in reviews. In this chapter, we propose a novel approach that combines the strengths of LLMs with conventional recommendation techniques by extracting the semantic aspects from user reviews. Although LLMs exhibit advanced proficiency across numerous tasks, they still encounter challenges in accurately identifying aspects, especially in autonomously discovering unseen aspects without explicit instructions. To mitigate this constraint of LLMs, we devise a chain-based prompting strategy, which consists of the *semantic aspects extraction prompt* and *semantic aspects-aware review extraction prompt*. The first prompt extracts semantic aspects from raw reviews, after which we refine and select high-quality aspects from the entire dataset. The second prompt then identifies reviews that align with these high-quality semantic aspects. On the basis of the extracted aspect-aware reviews, we can form aspect-specific interactions, capturing users' preferences from various semantic aspects. To demonstrate the potential of structuring raw reviews into semantic aspects via LLMs in recommendations, we present a **S**emantic **A**spect-based **G**raph **C**onvolution **N**etwork (SAGCN) model. This model performs graph convolution operations on aspect-aware interaction graphs, each of which is constructed based on the user-item

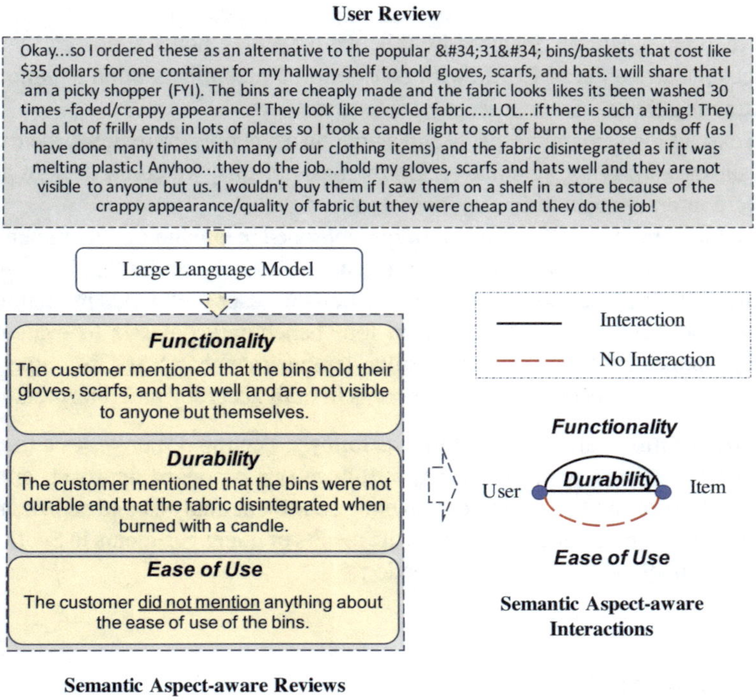

Fig. 4.1: An example of extracting semantic aspect-aware reviews (*e.g.*, functionality, durability, and ease of use) from user review via LLM. Based on the extracted reviews, the semantic aspect-aware interactions are discerned. The results from LLM indicate that user u interacts with item i in terms of functionality and durability aspects, but dose not interacts it in relation to ease of use.

interactions of a specific aspect derived from the aspect-aware reviews. The final user and item representations are obtained by combining the embeddings learned from different aspect-specific graphs. The merits of our model are twofold: (1) embeddings are enriched by learning from various aspects; (2) interpretability is improved by incorporating semantic aspects. Extensive experiments have been conducted on three benchmark datasets to evaluate the SAGCN. The results indicate that SAGCN outperforms state-of-the-art GCN-based recommendation models in generating more effective user and item representations. Additionally, interpretability analysis experiments demonstrate the effectiveness of integrating semantic aspects into our model. To facilitate further research, we have released the code and the constructed semantic aspect-aware interaction data for others to replicate this work [1].

In summary, the main contributions of this work are as follows:

[1] https://github.com/HuilinChenJN/LLMSAGCN.

- We emphasize the significance of utilizing semantic aspect-aware interaction in recommendations. Inspired by this, we propose a novel approach that combines the strengths of LLMs with conventional recommendation techniques by extracting the semantic aspects from user reviews.
- To extract the semantic aspects and semantic aspect-aware reviews, we devise a chain-based prompting strategy. On the basis of this, we could discern aspect-aware interactions within each user-item review.
- To demonstrate the effectiveness of our approach in enhancing recommendation accuracy and interoperability, we present the SAGCN model that leverages the extracted semantic aspect-aware interactions for user preference modeling.
- We conducted empirical studies on four benchmark datasets to evaluate the performance of SAGCN. Our results demonstrate that SAGCN outperforms state-of-the-art recommendation models in both accuracy and interpretability.

The rest of this chapter is structured as follows: Section 4.2 provides a thorough review of the related work. Next, we introduce our proposed approach and then provide details on its two main components: sentiment analysis and representation learning in Section 4.3. We then present the results of our experiments in Section 4.4. Finally, we conclude this chapter in Section 4.5.

4.2 Related Work

This section briefly introduces recent advances in collaborative filtering based on user reviews and large language models in recommendation, which are closely relevant to our work.

4.2.1 Collaborative Filtering Based on User Reviews

Collaborative Filtering (CF) techniques [76, 112, 135, 56, 136, 54, 86] play a pivotal role in recommendation systems. The fundamental concept involves learning representations (or embedding vectors) of users and items by harnessing collaborative information derived from user-item interactions.

To mitigate the data sparsity problem, user reviews [8, 103, 124, 21, 10, 22, 26] have been leveraged in recommender systems to learn the user and item representations as they can provide additional information about user preference. One of the typical methods to learn user preferences by exploiting reviews is through deep learning techniques [173]. For example, DeepCoNN [173] learns item characteristics and user preferences jointly from user reviews via the deep model. Existing methods extract aspects from reviews and integrate these aspects into user preference modeling. For example, A^{3}NCF [21] extracts user preferences and item characteristics from review texts via topic model to enhance the user's and item's representations. SULM [3] uses a sentiment analysis system, Opinion Parser, to extract aspects. Through a matrix factorization approach, it calculates the personal impact each aspect has on the user's overall rating. Furthermore, it is well-recognized that multimodal features can provide complementary information to each other, as demonstrated in [162, 168, 3, 86, 130]. JRL [168] first utilizes deep neural networks to separately extract user and item features from multiple modalities, such as ratings, reviews, and images, then forms the final user and item representations by concatenating those multimodal features. To model users' diverse preferences,

4.2.2 Large Language Models in Recommendation

Large Language Models (LLMs) have emerged as potent tools in natural language processing, garnering increasing attention in recommendation systems. The LLM-based models excel in capturing contextual information and comprehending user queries, item descriptions, and other textual data in a highly effective manner [45, 28, 120, 147]. By understanding the context, LLM-based models can improve the recommendation performance.

There are two primary approaches to integrating LLMs into recommendation systems. Firstly, LLMs can be adapted directly as recommendation models. Many works in this vein treat recommendation tasks as natural language tasks, employing techniques such as in-context learning, prompt tuning, and instruction tuning to tailor LLMs for generating recommendation results [65, 166, 164, 69, 79]. Recent attention has been particularly drawn to this line of research, buoyed by the impressive capabilities showcased by models like ChatGPT. For instance, Ji et al. [65] proposed a generative recommendation method that uses LLMs to generate recommended items. PEPLER [79] generates natural language explanations for recommendations by treating user and item IDs as prompts. Secondly, LLMs can be leveraged to extract high-quality representations of textual features or tap into the extensive external knowledge they encapsulate [94]. Most existing efforts align pre-trained models like BERT [31] with domain-specific data through fine-tuning. Additionally, beyond serving as standalone recommendation systems, some studies employ LLMs to construct model features. For example, GENRE [94] introduced three prompts to conduct three feature enhancement sub-tasks for news recommendation. The prompts involved refining news titles using ChatGPT, extracting profile keywords from user reading history, and generating synthetic news to enhance user interactions based on their history. By incorporating these LLM-constructed features, traditional news recommendation models can be greatly improved. RLMRec [110] generates user and item profiles based on interaction data and textual information, such as product descriptions, user reviews, and product titles. Then, these generated profiles are encoded as semantic information of LLMs and incorporated into the CF-based recommendation system.

Utilizing LLM's sentiment analysis capability, we aim to better understand users' various intentions expressed in reviews. Specifically, we first extract semantic aspects from user reviews and then identify the semantic-aspect interactions for each user and item pair. By leveraging semantic-aspect interactions, we can capture fine-grained user preferences. Compared to conventional methods that use features directly extracted from reviews by pretrained language models, semantic-aware interactions can improve the interpretability of recommendation models.

4.3 Methodology

4.3.1 Problem Setting and Notations Definition

To provide the details for our approach, let's first outline the problem setting. We are working with a dataset that consists of an interaction matrix denoted as $\boldsymbol{R}^{N \times M}$, where $\mathcal{U}$ represents the set of users, and $\mathcal{I}$ represents the set of items. N and M represent the total number of users and items, respectively. The matrix $\boldsymbol{R}$ records the interaction history between users and items. A nonzero entry $r_{ui} \in \boldsymbol{R}$ indicates that user $u \in \mathcal{U}$ has interacted with item $i \in \mathcal{I}$ before; otherwise, the entry is zero. For each user-item pair (u, i), a review v is associated with them. In this work, our goal is to explore the potential of reviews to enhance recommendation accuracy and interpretability.

To achieve this goal, we propose an approach that integrates LLMs with conventional recommendation techniques. Specifically, we (1) design a chain-based prompting approach, which can effectively extract semantic aspect-aware reviews without clear instructions; (2) introduce a semantic aspect-based GCN model for recommendation. We use $\{\boldsymbol{R}^{\mathbf{1}}, \boldsymbol{R}^{\mathbf{2}}, \cdots, \boldsymbol{R}^{\boldsymbol{A}}\}$ to denote the semantic aspect-based interaction matrices. A is the number of aspects. $\boldsymbol{R}^{\boldsymbol{a}}$ is the interaction matrix of the a-th aspect.

4.3.2 Overview

Recommendation systems learn user and item representations by exploiting user-item interactions, such as clicks and reviews. In this way, users' implicit preferences are encoded into latent vectors. Notably, user reviews offer explicit intents of a user towards an item. For example, the review might mention various semantic aspects such as *functionality*, *durability*, and *ease of use*. Each of these aspects carries a specific meaning related to user experience. Recognizing these explicit intentions leads to a deeper understanding of user preferences. More recently, LLMs have emerged as powerful tools for sentiment analysis [44, 120], capable of extracting semantic aspects from user reviews. Therefore, to enhance the accuracy of recommendation systems, a desired approach should well seamlessly combine the advanced capabilities of LLMs with conventional recommendation techniques.

In this chapter, we explore the fusion of the LLMs with conventional recommendation techniques to enhance both the accuracy and interoperability of recom-

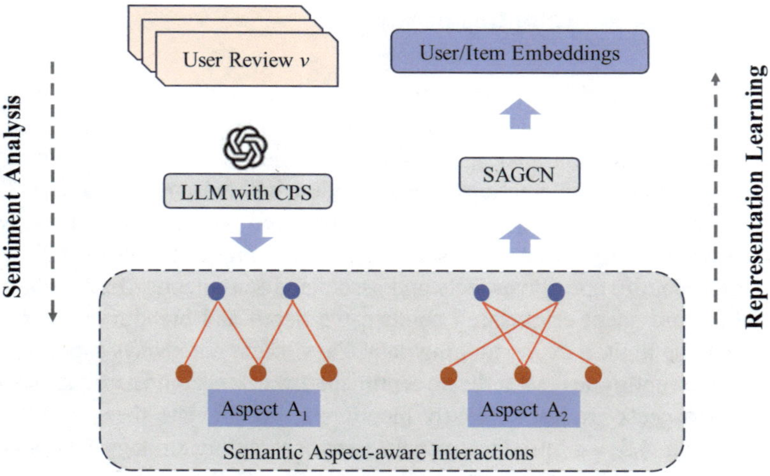

Fig. 4.2: Overview of Our Approach. Our approach consists of two components: sentiment analysis and representation learning. The former leverages the LLM with a Chain-based Prompting Strategy (short for CPS) to extract the semantic aspect-aware interactions from user reviews. The latter is dedicated to learning user and item representations based on these interactions, employing the Semantic Aspect-aware Graph Convolutional Network (short for SAGCN).

mendations. As shown in Fig. 4.2, our approach begins with the sentiment analysis of user-item reviews, utilizing LLMs with the chain-based prompting strategy to extract semantic aspect-aware interactions from each review. These extracted semantic aspect-aware reviews shed light on the fine-grained interactions underlying each user-item relationship. On the basis of these insights, we introduce a semantic aspect-based graph convolution network to enhance the user and item representations by leveraging these semantic aspect-aware interactions. To this end, our approach can well extract and leverage the semantic aspect-aware interactions among each user-item pair for better user and item representation learning.

Next, we detail our method for semantic aspect-aware review extraction and semantic aspect-based graph convolution network.

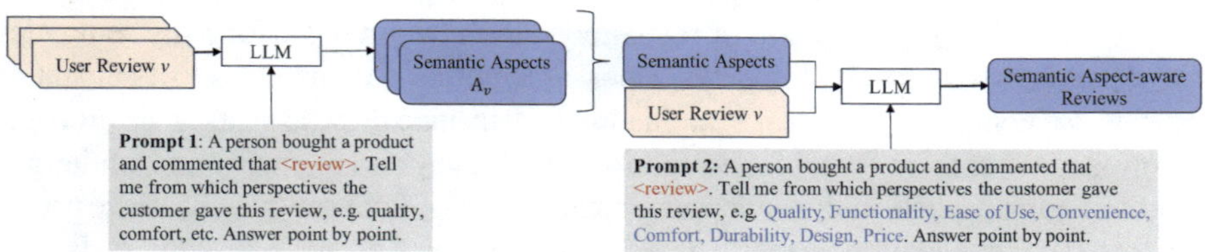

(a) Semantic Aspects Extraction (b) Semantic Aspects-aware Review Extraction

Fig. 4.3: Semantic Aspects and Semantic Aspect-aware Reviews Extraction.

4.3.3 Chain-based Prompting Strategy

In this section, we conduct sentiment analysis on user reviews using LLM to discern user-item interactions on various semantic aspects. Despite the powerful capability of LLMs in sentiment analysis, it is still a challenge to accurately identify the semantic aspect-aware interactions in each review without any instruction. This is because user reviews often contain complex, nuanced language that varies widely in terms of vocabulary, syntax, and semantics. This complexity makes it difficult to consistently identify specific aspects and associated sentiments. On the other side, while LLMs are adept at grasping context, the depth and breadth of their understanding can be limited by the training data. They might not always capture the full context or the subtle cues that indicate sentiment towards specific aspects, especially when these aspects are not explicitly mentioned. To alleviate these problems, as depicted in Fig. 4.3, we introduce a chain-based prompting strategy that begins by collecting the semantic aspects from all user reviews. Subsequently, we refine and select high-quality semantic aspects from all reviews and use these as the instruction information to generate the semantic aspect-aware reviews for each review. Finally, we identify aspect-aware interactions based on the generated reviews.

Semantic Aspects Extraction

The semantic aspects vary depending on the recommendation scenario. For instance, when buying dresses, *style* is crucial, but for office products, *ease of use* is the main concern. In this work, we first utilize the understanding capabilities of LLMs to extract unseen semantic aspects from the dataset. As illustrated in Fig. 4.3(a), the semantic aspects extraction prompt is designed as follows:

- *Prompt 1*: *A person bought a product and commented that <review>. Tell me from which perspectives the customer gave this review, e.g., quality, comfort, etc. Answer point by point.*

Where $<$ review $>$ denotes a user review, let the review v be analyzed by a LLM to generate a set of semantic aspects $\mathcal{A}_v = \{a_v^1, a_v^2, \ldots, a_v^{N_v}\}$, where N_v is the number of extracted aspects. It is important to note that the prompt provides minimal information about these semantic aspects. We utilize *quality* and *comfort* as indicative cues for the LLM to identify the unseen semantic aspects within the reviews. Through the analysis of each review by the LLM, we can compile a comprehensive list of semantic aspects pertinent to the domain of recommendation by amalgamating the semantic aspect sets derived from all reviews. Upon assembling a complete set of semantic aspects, we conduct a statistical examination to prioritize these aspects. This process involves ranking them from highest to lowest frequency and manually consolidating or eliminating aspects with overlapping meanings. The final set of semantic aspects for this recommendation scenario is denoted as $\mathcal{A} = \{a^1, a^2, \ldots, a^N\}$, N is the number of obtained semantic aspects.

Semantic Aspect-Aware Review Extraction

Empirically, LLMs tend to generate more accurate results when provided with detailed prompts. To accurately identify the semantic aspects within each review, we further employ LLM to generate high-quality semantic aspect-aware reviews based on the obtained semantic aspect set $\mathcal{A}$. As shown in Fig. 4.1, an aspect-aware review provides a summary that identifies the presence of specific aspects within a review. Since we can focus on specific aspects, high-quality semantic aspect-aware reviews can be obtained. For instance, on the Baby Dataset, we get a semantic aspect set containing *Quality*, *Comfort*, *Durability*, *Design*, *Functionality*, *Ease of use*, *Price*, and *Size*. As illustrated in Fig. 4.3(b), we adapt the previous prompt to specific semantic aspects:

- *Prompt 2*: *A person bought a product and commented that <review>. Tell me from which perspectives the customer gave this review, e.g.,* $\{a^1, a^2, \cdots, a^N\}$*. Answer point by point.*

The LLM identifies the semantic aspect-aware review for the aspects we input. Then, we identify the semantic aspect-aware interactions between the user and the item based on the generated semantic-aware reviews. For example, if the LLM generates the output: "*The customer mentioned that the bins were not durable and that the fabric disintegrated when burned with a candle.*", it indicates an interaction on *Functionality*. On the other hand, if the LLM generates the output: "*The customer did not mention anything about the ease of use of the bins.*", which means that there is no interaction in this aspect. Note that the time costs for the Chain-based Prompting strategy vary based on the length of input and output tokens. Based on our empirical evaluations, the process takes approximately 4 to 10 seconds per review.

4.3.4 Semantic Aspect-Based Graph Convolution Network

Based on the extracted semantic-aware user-item interactions for semantic aspects, we seek to model users' preferences on various aspects with GCN-based techniques. The structure of our SAGCN is illustrated in Fig. 4.4, and it is composed of three key components: 1) **embedding initialization**, which initialized node (users and items) embeddings for the subsequent learning in each aspect-specific interaction graph; 2) **embedding propagation**, in which the GCN technique [54] is adopted to learn the user and item embeddings by performing graph convolution on each semantic aspect-based graph; 3) **embedding generation**, which aggregates the embeddings learned from each aspect-specific graph and combines the embeddings of different aspects.

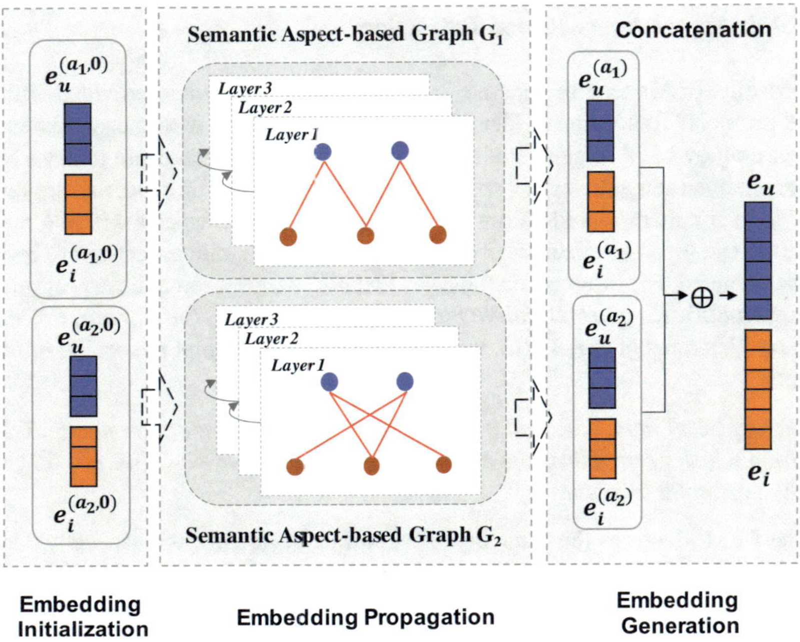

Fig. 4.4: Overview of our SAGCN model.

Semantic Aspect-based Graph Construction

The semantic aspect-based graph construction is built on the results of aspect-aware review extraction, which is achieved by two steps with a designed chain-based prompting approach. We extract semantic aspects using the *semantic aspects extraction prompt* and then extract aspect-aware reviews based on those aspects using the *semantic aspects-aware review extraction prompt*.

Semantic aspect-aware interaction: if a user u's review contains comments of a specific on a certain aspect of item i, then we regard there is an interaction between u and i on this aspect. By identifying the aspects of user reviews on each item, we can form semantic aspect-aware interaction graphs to exploit the semantic aspects in user and item representation learning.

In previous studies [136, 54], the user-item interaction graph is directly built based on the users' historical interactions. Specifically, let $\mathcal{G} = (\mathcal{W}, \mathcal{E})$ be an undirected graph, where $\mathcal{W}$ denotes the set of nodes and $\mathcal{E}$ is the set of edges. Specifically, $\mathcal{W}$ consists of two types of nodes: users $u_i \in \mathcal{U}$ with $i \in \{1, \cdots, N_u\}$, items $v_j \in \mathcal{V}$ with $j \in \{1, \cdots, N_v\}$, and $\mathcal{U} \cup \mathcal{V} = \mathcal{W}$. By extracting the semantic aspect-aware interactions from reviews using LLM, we separately construct the semantic aspect-specific graphs based on the user-item interactions on various aspects. In this work, we defined $\mathcal{G}_a = (\mathcal{W}, \mathcal{E}_a)$ as the semantic aspect-based graph for aspect a, where $\mathcal{E}_a$

is the set of edges.

Embedding Initialization

In the proposed SAGCN, the vectors of users and items are composed of embeddings from A aspects. Following existing GCN-based recommendation models [136, 54, 87], we initialize the embedding vectors for each aspect. As an example, the ID embeddings of user $u \in \mathcal{U}$ for different aspects are initialized as:

$$\boldsymbol{e}_{\boldsymbol{u}}^{(0)} = \{\boldsymbol{e}_{\boldsymbol{u}}^{(1,0)}, \boldsymbol{e}_{\boldsymbol{u}}^{(2,0)}, \cdots, \boldsymbol{e}_{\boldsymbol{u}}^{(A,0)}\}, \tag{4.1}$$

where $\boldsymbol{e}_{\boldsymbol{u}}^{(0)} \in \mathbb{R}^{d \times A}$ is ID embedding capturing intrinsic characteristics of user u; $\boldsymbol{e}_{\boldsymbol{u}}^{(a,0)} \in \mathbb{R}^{d}$ is u's representation in aspect a. Analogously, $\boldsymbol{e}_{\boldsymbol{i}}^{(0)} = \{\boldsymbol{e}_{\boldsymbol{i}}^{(1,0)}, \boldsymbol{e}_{\boldsymbol{i}}^{(2,0)}, \cdots, \boldsymbol{e}_{\boldsymbol{i}}^{(A,0)}\}$ is defined as the ID embeddings of item $i \in \mathcal{I}$. In addition, it is worth noting that our model randomly and separately initializes the embeddings of the user IDs and item IDs for various aspects.

Embedding Propagation

To exploit the high-order connectives on each aspect to learn user and item representations, we propagate embeddings across various semantic aspect-based graphs. Take the semantic aspect-specific graph $\mathcal{G}_a$ as an example, after propagating embeddings through k graph convolution layers, we define $\boldsymbol{e}_{\boldsymbol{u}}^{(a,k)}$ and $\boldsymbol{e}_{\boldsymbol{i}}^{(a,k)}$ to denote the embedding of user u and item i, which are propagated from $k-$th order neighboring nodes, where $k \geq 0$, the embedding propagation is formulated as:

$$\begin{aligned} \boldsymbol{e}_{\boldsymbol{u}}^{(a,k+1)} &= \sum_{i \in \mathcal{N}_u^a} \frac{1}{\sqrt{|\mathcal{N}_u^a|}\sqrt{|\mathcal{N}_i^a|}} \boldsymbol{e}_{\boldsymbol{i}}^{(a,k)}, \\ \boldsymbol{e}_{\boldsymbol{i}}^{(a,k+1)} &= \sum_{u \in \mathcal{N}_i^a} \frac{1}{\sqrt{|\mathcal{N}_i^a|}\sqrt{|\mathcal{N}_u^a|}} \boldsymbol{e}_{\boldsymbol{u}}^{(a,k)}, \end{aligned} \tag{4.2}$$

where $\boldsymbol{e}_{\boldsymbol{i}}^{(a,k+1)}$ and $\boldsymbol{e}_{\boldsymbol{u}}^{(a,k+1)}$ can be considered as embeddings propagated from the $(k+1)$-th order neighboring nodes. $\frac{1}{\sqrt{|\mathcal{N}_u^a|}\sqrt{|\mathcal{N}_i^a|}}$ is symmetric normalization terms; $\mathcal{N}_u^a$ denotes the set of items that interact with user u, and $\mathcal{N}_i^a$ denotes the set of users that interact with item i.

Embedding Combination

After the K-layers graph convolution, the embeddings of user u and item i of aspect a are aggregated from all layers:

$$\boldsymbol{e}_{\boldsymbol{u}}^{(a)} = \sum_{k=0}^{K} \boldsymbol{e}_{\boldsymbol{u}}^{(a,k)}, \boldsymbol{e}_{\boldsymbol{i}}^{(a)} = \sum_{k=0}^{K} \boldsymbol{e}_{\boldsymbol{i}}^{(a,k)}. \tag{4.3}$$

The final representations of user u and item i are a concatenation of their embeddings learning in multiple semantic aspect-based graphs:

$$\boldsymbol{e_u} = \boldsymbol{e_u^{(0)}}||\cdots||\boldsymbol{e_u^{(a)}}||\cdots||\boldsymbol{e_u^{(A)}}, \boldsymbol{e_i} = \boldsymbol{e_i^{(0)}}||\cdots||\boldsymbol{e_i^{(a)}}||\cdots||\boldsymbol{e_i^{(A)}}, \quad (4.4)$$

where $\boldsymbol{e_u^{(a)}}$ and $\boldsymbol{e_i^{(a)}}$ denote the embeddings of users and items by performing graph convolution on graph $\mathcal{G}_a$, respectively.

Model Training

After learning the representation for both users and items, the preference of a given user u for a target item i is estimated as follows:

$$\hat{r}_{ui} = \boldsymbol{e_u^T e_i}. \quad (4.5)$$

In this work, we aims to recommend a list of items to the target users by computing their preference score. For the optimization, we utilize the pairwise learning method, which is similar to the approach employed in previous works [54, 87]. To train our SAGCN model, we construct a triplet of $\{u, i^+, i^-\}$, comprising a positive item (an observed interaction between u and i^+) and a negative item (an unobserved interaction between u and i^-). The objective function is formulated as follows:

$$\arg\min \sum_{(\mathbf{u},\mathbf{i}^+,\mathbf{i}^-)\in\mathcal{O}} -\ln\phi(\hat{r}_{ui^+} - \hat{r}_{ui^-}) + \lambda\,\|\Theta\|_2^2 \quad (4.6)$$

where $\mathcal{O}$ represent the training set, defined as $\mathcal{O} = \{(u, i^+, i^-)|(u, i^+) \in \mathcal{R}^+, (u, i^-) \in \mathcal{R}^-\}$. Here, $\mathcal{R}^+$ indicates the observed interactions in the training dataset, while $\mathcal{R}^-$ denotes a set of sampled unobserved interactions. Additionally, λ represents the regularization weight, and Θ stands for the model parameters. We employ L_2 regularization to mitigate overfitting. To optimize our model and update its parameters, we utilize the mini-batch Adam technique, as described in [70].

4.4 Experiments

To validate the effectiveness of our model, we conducted extensive experiments on four public datasets to answer the following research questions:

RQ1: Does our SAGCN model outperform state-of-the-art methods on the top-n recommendation task?

RQ2: Does our SAGCN model outperform the pure LLM on the top-n recommendation task?

RQ3: How does the number of graph convolution layers affect the performance of SAGCN?

RQ4: How does our model handle noisy neighboring nodes?

Table 4.1: Statistics of the Experimental Datasets.

Dataset	#user	#item	#interactions	sparsity
Office	4,905	2,420	53,258	99.55%
Goodreads	4,545	5,274	53,458	99.78%
Baby	19,445	7,050	160,792	99.88%
Clothing	39,387	23,033	278,677	99.97%

RQ5: How do the embedding dimensions affect the performance of our SAGCN model?

RQ6: Can our model effectively extract the semantic aspects? How does the number of aspects affect the performance of recommendation?

RQ7: How does our method provide interpretability of the recommendation results?

RQ8: Are the features associated with each pair of semantic aspects mutually independent?

In the next, we first introduced the experimental setup, and then answered the above research questions in sequence.

4.4.1 Experimental setup

Data Description

To evaluate the effectiveness of our approach, we conducted experiments on Amazon Product Datasets and Goodreads Review Datasets.

Amazon Product Datasets [2]. These datasets stem from the extensive product review system of Amazon. It includes user-product interactions, reviews, and metadata. In our experiments, we focused solely on user-product interactions and user reviews. Three product categories from the 5-core version of these datasets are selected for our evaluation: Office Products (Office), Baby, and Clothing. This indicates that both users and items have a minimum of 5 interactions.

Goodreads Review Datasets [3]. The datasets stem from the product review of GoodReads, which is the largest site for readers and book recommendations. It collected user-book interactions, reviews, and book metadata with anonymized user IDs and book IDs. In our experiments, we only use user-book interactions and review text.

In line with previous studies on conventional recommendation methods [54, 87], we randomly split each user's interaction into training and testing sets with a ratio of 80:20 for each dataset. Within the training set, we randomly select 10% of interac-

[2] http://jmcauley.ucsd.edu/data/amazon

[3] https://www.goodreads.com/

tions to establish the validation set for hyper-parameter optimization. The pairwise learning strategy is adopted for parameter optimization. Specifically, the observed user-item interactions in the training set were treated as positive instances. For each positive instance, we paired it with a random sampling of negative instances that the user had not interacted with yet. For a fair comparison, we consistently adopt the negative sampling strategy across our methods and competitors. To further validate the effectiveness of our approach, we compare our proposed SAGCN with a pure LLM-based method, and the details of test dataset construction for the LLM-based method can be seen in Section 4.4.3.

Compared Baselines

To demonstrate the effectiveness of our approach, we compared our SAGCN with several state-of-the-art models, which include CF-based recommendation methods without side information (*i.e.*, NeuMF [4], GCMC [5], NGCF [6], LightGCN [7], DGCF [8], IMP-GCN [9], NCL [10]), multimodal recommendation methods (*i.e.*, MMGCN [11], GRCN [12], LATTICE [13], BM3 [14]) and review-based recommendation methods (*i.e.*, DeepCoNN [15], NARRE [16], RGCL [17], LightGCN$_{LDA}$). For fair comparisons, we only adopt reviews as the side information for MMGCN, GRCN, LATTICE, and BM3. In addition, we devised a LLM-based recommendation method and adopted two LLM-based recommendation methods (*i.e.*, LLamaRec [18] and RLMRec [19]) for comparison, the details are provided in Section 4.4.3.

- **NeuMF [56]**: This is a state-of-the-art neural collaborative filtering method that present a general framework for collaborative filtering based on neural networks. It innovatively introduces multiple hidden layers to capture the non-linear interactions between the latent features of users and items.
- **GCMC [5]**: It is a graph-based auto-encoder framework for matrix completion. This method leverages the GCN technique [71] on the user-item interaction graph

---

[4] https://github.com/hexiangnan/neural_collaborative_filtering

[5] https://github.com/riannevdberg/gc-mc

[6] https://github.com/xiangwang1223/neural_graph_collaborative_filtering

[7] https://github.com/kuandeng/LightGCN

[8] https://github.com/xiangwang1223/disentangled_graph_collaborative_filtering

[9] https://github.com/liufancs/IMP_GCN

[10] https://github.com/RUCAIBox/NCL

[11] https://github.com/weiyinwei/MMGCN

[12] https://github.com/weiyinwei/GRCN

[13] https://github.com/CRIPAC-DIG/LATTICE

[14] https://github.com/enoche/BM3

[15] https://github.com/chenchongthu/DeepCoNN

[16] https://github.com/chenchongthu/NARRE

[17] https://github.com/JarenceSJ/ReviewGraph

[18] https://github.com/Yueeeeeeee/LlamaRec

[19] https://github.com/HKUDS/RLMRec

and solely uses one convolutional layer to exploit direct connections between the nodes of users and items.

- **NGCF [136]**: This is the very first GCN-based recommendation model that incorporates high-order connectivity of user-item interactions. It encodes the collaborative signal from high-order neighbors by performing embedding propagation on the user-item bipartite graph.
- **LightGCN [54]**: It is a state-of-the-art graph collaborative filtering method. This model can be considered a simplified version of NGCF [136], where certain components such as the feature transformation and non-linear activation module are removed.
- **IMP-GCN [87]**: This method performs high-order graph convolution within sub-graphs, each of which is constructed based on users with shared interests and associated items, thus avoiding the negative information passed from high-order neighbors.
- **NCL [82]**: This is a model-agnostic contrastive learning framework, which explicitly incorporates the potential neighbors into contrastive pairs. It aims to capture the correlation between a node and its prototype, thereby refining the neural graph collaborative filtering techniques.
- **LATTICE [167]**: This method uses side information to construct an item-item graph via a modality-aware structure learning layer and perform graph convolution operations to aggregate information from high-order neighbors to enhance the item representations.
- **MMGCN [140]**: The model learns user preferences and item characteristics by propagating modality embeddings on the user-item interaction graph.
- **GRCN [143]**: This GCN-based multimedia recommendation model designs a graph refining layer to adjust the structure of the interaction graph adaptively based on the status of model training for discovering and pruning potential false-positive edges.
- **BM3 [175]**: This self-supervised framework uses a latent representation dropout mechanism to generate the target view of a user or an item. Moreover, it is trained without negative samples by jointly optimizing three multi-modal objectives.
- **DeepCoNN [173]**: This model jointly learns item properties and user behavior from review text by using Convolutional Neural Networks (CNNs) to encode the textual data.
- **NARRE [10]**: The model uses two parallel CNNs to learn the latent representation of reviews. Then, an attention mechanism is designed to estimate the contributions of different reviews for the item and user.
- **RGCL [118]**: This model constructs a review-aware user-item graph with feature-enhanced edges derived from reviews. Each edge feature incorporates both the user-item rating and the semantic information extracted from the corresponding review.
- **LightGCN$_{LDA}$ [6]**: To validate the ability of LLMs in understanding reviews, we designed a baseline that combines LDA [20] with LightGCN. LDA extracts

[20] https://github.com/lda-project/lda

topics from the dataset, and we calculate the similarity between each review and these topics. Using these similarities, we construct topic-specific graphs, where LightGCN learns user and item representations through graph convolutions.

- **LlamaRec [161]**: The model is a two-stage ranking-based recommendation. This method first retrieves a series of candidate items by small-scale recommenders learning based on the user interaction history. Then, the model adopts an LLM-based ranking model to reranking the order of candidate items.
- **RLMRec [110]**: The model designs an LLM-based generation framework that generates user and item profiles based on interaction data and textual information, such as product descriptions, user reviews, and product titles. Then, these generated profiles are encoded as semantic information of LLMs and incorporated into the CF-based recommendation system.

Evaluation Metrics

Two widely-used evaluation metrics are employed for evaluating top-K recommendations: *Recall* and *Normalized Discounted Cumulative Gain* (NDCG) [52]. For conventional recommendation methods, the recommendation accuracy is calculated for each metric based on the top 10 and 20 results. The reported results are the average values across all testing users. In contrast, for the LLM-based method we have devised for comparison, we calculate the recommendation accuracy based on the top 1, 3, 5, and 7 results. In addition, the reported results are the average values across the sampled users. Note that we sample several users for evaluation due to the high resource and time consumption of LLMs.

Parameter Settings

The PyTorch framework [21] is adopted to implement the proposed model. We employed Vicuna-13B [22] to analyze the user reviews. It has been fine-tuned from LLaMA using supervised instruction-based fine-tuning and exhibits strong performance in sentiment analysis. In our experiments, all hyperparameters are carefully tuned. The user and item embedding size for all competitor methods is set to 64 across all datasets. The number of semantic aspects N is defined as 8 across all four datasets. The embedding size associated with each aspect is also set to 64 within our proposed SAGCN. The mini-batch size is fixed to 1024. The learning rate for the optimizer is searched from $\{1e^{-5}, 1e^{-4}, \cdots, 1e^{1}\}$, and the model weight decay is searched in the range $\{1e^{-5}, 1e^{-4}, \cdots, 1e^{-1}\}$. Besides, the early stopping strategy is implemented to enhance training efficiency. Specifically, the training process will stop if recall@10 does not increase for 20 successive epochs.

[21] https://pytorch.org.

[22] https://huggingface.co/lmsys/vicuna-13b-v1.5

Table 4.2: Performance of our method and the competitors on the Office dataset and the Baby dataset.

Datasets	Office				Baby			
Metrics	R@10	N@10	R@20	N@20	R@10	N@10	R@20	N@20
NeuMF	5.14	3.89	8.12	5.21	3.11	2.11	4.85	2.69
GCMC	6.72	5.27	10.27	6.79	4.55	2.99	7.24	3.89
LightGCN	9.87	6.04	14.47	7.43	5.94	3.30	9.25	4.20
DGCF	9.95	6.25	14.37	7.52	5.90	3.27	9.20	4.14
IMP-GCN	10.11	6.36	14.47	7.71	6.24	3.49	**9.56**	4.38
NCL	10.07	6.30	14.40	7.65	6.15	3.42	9.45	4.30
MMGCN	5.74	3.42	9.39	4.54	3.95	2.17	6.46	2.85
GRCN	**10.38**	6.34	**15.33**	**7.81**	5.57	3.03	22.64	12.67
LATTICE	10.00	6.09	14.99	7.57	6.06	3.40	9.29	4.27
BM3	9.80	6.09	14.02	7.36	**6.44**	**3.65**	9.52	**4.48**
DeepCoNN	5.32	4.01	8.35	5.33	3.20	2.02	5.05	2.71
NARRE	6.12	4.78	9.41	6.15	4.02	2.32	6.14	2.98
RGCL	7.89	5.69	12.40	7.02	5.22	2.54	8.20	3.19
LightGCN$_{LDA}$	10.12	6.24	15.12	7.67	5.98	3.44	9.41	4.29
LlamaRec	9.89	5.83	14.44	7.20	5.62	3.11	8.83	3.85
RLMRec	10.1	**6.35**	14.87	7.78	5.84	3.37	8.91	4.30
SAGCN	**11.71***	**7.34***	**16.71***	**8.84***	**7.35***	**4.23***	**10.56***	**5.09***
Imporv.	12.81%	15.59%	9.00%	13.19%	14.13%	15.89%	10.46%	13.61%

The symbol * denotes that the improvement is significant with $p - value < 0.05$ based on a two-tailed paired t-test. Notice that the values are reported by percentage with '%' omitted.

4.4.2 Performance Comparison With Baseline Methods (RQ1)

In this section, we compared the performance of our proposed SAGCN with all the adopted competitors (conventional recommendation models). The experimental results of our method and all compared methods over four test datasets in terms of Recall@10, Recall@20, NDCG@10, and NDCG@20 are reported in Table 4.2 and Table 4.3. Note that the best and second-best results are highlighted in bold. From the results, we have the following observations.

We first focus on the performance of methods in the first block of both Table 4.2 and Table 4.3, which only uses the user-item interaction information. NeuMF is a deep learning-based method, and it shows relatively poor performance. GCMC performs better than NeuMF, highlighting the benefits of leveraging graph structure. However, GCMC underperforms LightGCN because it not explicitly leverage high-order connectivities between users and items. This showcases the strength of GCN and the importance of incorporating high-order information into representation learning. Moreover, DGCF performs better than LightGCN by leveraging the

Table 4.3: Performance of our method and the competitors on the Clothing dataset and the Goodreads dataset.

Datasets	Clothing				Goodreads			
Metrics	R@10	N@10	R@20	N@20	R@10	N@10	R@20	N@20
NeuMF	0.94	0.54	1.50	0.71	9.24	7.45	14.63	9.02
GCMC	3.17	1.86	4.86	2.35	16.32	10.58	22.32	12.32
LightGCN	4.45	2.43	6.44	2.95	16.99	11.02	23.35	13.01
DGCF	4.67	2.68	6.91	3.32	16.89	11.22	21.76	13.12
IMP-GCN	4.80	2.76	7.11	3.40	18.05	12.07	23.79	14.12
NCL	4.76	2.74	7.10	3.37	17.69	11.64	23.55	13.68
MMGCN	2.42	1.29	3.76	1.64	11.24	7.85	15.33	9.48
GRCN	4.47	2.35	6.70	2.94	16.62	10.79	22.64	12.67
LATTICE	5.03	2.79	7.28	3.37	17.92	11.35	23.60	13.27
BM3	**5.28**	**2.94**	**7.75**	**3.58**	**18.28**	**12.12**	**24.68**	**14.15**
DeepCoNN	1.89	1.01	2.98	1.35	11.89	7.22	5.66	8.44
NARRE	2.37	1.32	3.62	1.69	12.79	9.32	16.84	10.89
RGCL	3.32	1.93	5.22	2.41	16.55	10.43	21.79	12.19
LightGCN$_{LDA}$	5.17	2.88	7.50	3.44	16.73	10.87	22.07	12.70
LlamaRec	3.72	2.17	5.45	2.68	16.67	10.82	21.80	12.66
RLMRec	4.48	2.60	6.65	3.08	16.72	11.02	22.02	12.71
SAGCN	**6.07***	**3.58***	**8.44***	**4.20***	**19.40***	**13.15***	**26.17***	**15.14***
Imporv.	14.96%	21.77%	8.90%	17.32%	6.15%	8.50%	6.03%	7.03%

The symbol * denotes that the improvement is significant with $p - value < 0.05$ based on a two-tailed paired t-test. Notice that the values are reported by percentage with '%' omitted.

disentangled representation technique to model users' diverse intents. Both IMP-GCN and NCL consistently outperform all the above GCN-based methods over all the datasets. This is mainly due to their ability to filter out noisy information from neighboring nodes effectively. In particular, IMP-GCN's performance superiority over LightGCN and DGCF demonstrates the importance of distinguishing nodes in high-order neighbors in the graph convolution operation.

The second block of both Table 4.2 and Table 4.3 is the GCN-based methods utilizing user reviews as side information. MMGCN is the first method combining side information with GCN, it exploits the high-order information of various modalities. However, it falls short of GRCN in terms of effectiveness. The reason is that the GRCN could adaptively adjusting the false-positive edges of the interaction graph to discharge the ability of GCN. LATTICE outperforms GRCN on Baby and Clothing by constructing item-item graphs based on item similarity in the side information. BM3 consistently performs better than all general CF baselines over Baby and Clothing. It should be credited to the joint optimization of multiple objectives to learn the representations of users and items.

The third block of both Table 4.2 and Table 4.3 illustrates that review-based methods primarily use user textual information as a key semantic feature. DeepCoNN, which employs Convolutional Neural Networks, designs two coupled networks to learn user and item representations from reviews, outperforming NeuMF due to its use of semantic information. NARRE enhances recommendations by using attention mechanisms to weigh the significance of different reviews, showing particular improvement on the sparse Clothing dataset. RGCL further incorporates review data by building a review-aware user-item graph, outperforming all three review-based methods. To validate the ability of LLMs in understanding reviews, we designed LightGCN$_{LDA}$, which combines LDA with LightGCN. Similar to our approach, it uses topic-level information to construct user-item graphs and then learns user and item representations through GCN techniques. As a result, it outperforms all review-based models across all datasets.

The fourth block of both Table 4.2 and Table 4.3 consists of two LLM-based methods: LlamaRec and RLMRec. LlamaRec is a two-stage framework that first retrieves candidate items and then ranks them using an LLM. It outperforms NeuMF due to the advanced capabilities of LLMs in understanding complex patterns and relationships within the data. RLMRec, on the other hand, outperforms LlamaRec by leveraging a more comprehensive data approach and generating richer user and item profiles. While LlamaRec focuses solely on user interaction data, RLMRec combines interaction data with textual information such as product descriptions, user reviews, and product titles. This enables RLMRec to create detailed, contextually enriched profiles, offering a deeper understanding of user preferences and item characteristics.

SAGCN consistently yields the best performance across all datasets. In particular, the improvements of SAGCN over the strongest competitors $w.r.t$ NDCG@10 are 15.59%, 15.89%, 21.77%, and 8.50% in Office, Baby, Clothing, and Goodreads, respectively. We credit this to the combined effects of the following three aspects. Firstly, we accurately extract structured semantic aspect-aware reviews using LLM to understand user behaviors at a finer level, which also avoids the noise in raw reviews. Our proposed method outperforms LightGCN$_{LDA}$, highlighting the superior understanding capabilities of LLMs. Additionally, the semantic aspects generated by LLMs offer better interpretability compared to the latent topics extracted by LDA. Secondly, the SAGCN performs graph convolution on various semantic-aspect-based graphs, which can alleviate the over-smoothing problem [87] and avoid negative information from unreliable neighbor nodes [143]. Thirdly, we extract fine-grained user behaviors for each semantic aspect to bridge the modality gap between interaction and textual features. Different with existing methods that integrate textual features with user-item interaction data for representation learning, our approach constructs structured user-item interactions from user reviews via LLM, which are associated with various semantic aspects. These structured interactions are then leveraged for the representation learning of users and items.

4.4.3 Performance Comparison with Pure LLM (RQ2)

To further validate the effectiveness of our approach, we implemented a method named **LLM***, built upon LLM for comparative analysis. However, due to the high resource and time consumption of LLM, we evaluated the method by randomly selecting 200 users from each dataset. To construct the testing data, we randomly selected one positive item from the history items of the original testing data. We then sampled 9 negative items not interacted with by the user as the testing set for each user. Next, we converted the historical data and items in the testing set for each user into textual format using a prompt template. Finally, we fed the converted text into LLM to generate scores for the items in the testing set. These items are then ranked according to the scores generated by the LLM. Our prompt template is as follows:

I want you to rate every candidate product's historical record of purchased habits. You are encouraged to learn his preferences from the historical records he has purchased. Here are The historical interactions of a user include: { history }. Now, how will the user rate these candidate products? (1 being lowest and 5 being highest) { candidates }. Importantly, the interacted items should have been excluded from the rating. Finally, Only output rating item list, which template is: 1. Swingline GBC UltraClear Thermal Laminating Pouches, Menu Size, 3 Mil, 25 Pack (item id: B00006IA2K) - Rating: 4.0 stars

Where $\{history\}$ consists of item title and user rating score; $\{candidates\}$ is composed of item titles and item id. Besides, we also use the **Random** method as a comparison method, which randomly ranks the items in the testing set.

Table 4.4: Performance Comparison of Random, LLM* and our method on Office.

Datasets	Office							
Metrics	k=1		k=3		k=5		k=7	
	Recall	NDCG	Recall	NDCG	Recall	NDCG	Recall	NDCG
Random	0.0835	0.0835	0.2926	0.2012	0.5261	0.2962	0.7203	0.3632
LLM*	0.1086	0.1086	0.5803	0.3903	0.7660	0.4657	0.9122	0.5163
SAGCN	0.4880	0.4880	0.7470	0.6371	0.8763	0.6838	0.9669	0.7289

Table 4.5: Performance Comparison of Random, LLM* and our method on Clothing.

Datasets	Clothing							
Metrics	k=1		k=3		k=5		k=7	
	Recall	NDCG	Recall	NDCG	Recall	NDCG	Recall	NDCG
Random	0.0841	0.0841	0.2848	0.2208	0.4474	0.2867	0.6186	0.3458
LLM*	0.0912	0.0912	0.4860	0.2997	0.7692	0.4170	0.8913	0.4669
SAGCN	0.4823	0.4823	0.7239	0.6176	0.8273	0.6603	0.9245	0.6908

Table 4.6: Performance Comparison of Random, LLM* and our method on Baby.

Datasets	Baby							
Metrics	k=1		k=3		k=5		k=7	
	Recall	NDCG	Recall	NDCG	Recall	NDCG	Recall	NDCG
Random	0.0537	0.0537	0.3154	0.2166	0.4899	0.2888	0.6913	0.3579
LLM*	0.0738	0.0738	0.4094	0.2605	0.7384	0.3954	0.9463	0.4681
SAGCN	0.3841	0.3841	0.7970	0.6899	0.8993	0.6913	0.9812	0.7293

As illustrated in Tables 4.4, 4.5 and 4.6 the performance of the Random method is inferior as it randomly recommends the items to users. The LLM* method achieves better performance than Random. This is because it has the ability to recognize patterns and correlations in data. This capability allows it to predict user preferences and make more informed recommendations based on observed patterns in user behavior or item characteristics. However, its performance does not significantly exceed that of Random, especially in terms of Recall@1 and NDCG@1. This can be attributed to the LLM not having been trained or fine-tuned for the specific recommendation task. Experimental results demonstrate the effectiveness of our proposed approach, which integrates LLM with conventional recommendation techniques. A remarkable advantage of SAGCN is its aspect-aware graph, which allows for a finer understanding of complex interactions between users and items, further augmented by the sentiment analytical power of LLM.

4.4.4 Effect of Layer Numbers (RQ3)

In this section, we evaluated the performance of our model by varying the number of layers in the graph convolution process. Our primary goal was to examine whether our model can mitigate the over-smoothing problem often encountered in graph convolutions. To investigate the effectiveness of our model, we iteratively increased the number of convolution layers from 1 to 7. We then compared its performance against four GCN-based methods: LightGCN, DGCF, IMP-GCN, GRCN, and BM3. The comparative outcomes are depicted in Fig. 4.5 and 4.6.

From the results, all three models (DGCF, NCL, and BM3) achieve their peak performance when stacking 2 or 3 layers across two datasets. Beyond this, they will cause dramatic performance degradation with increasing layers. These results indicate that these three models suffer from the over-smoothing problem and inadvertently introduce noise information in the deep structure. GRCN achieves the peak performance when stacking 4 layers. This is attributed to discovering and pruning the potential false-positive edges, which avoid the noisy information from unreliable neighboring nodes. IMP-GCN can achieve robust performance with a deeper structure. This resilience can be credited to its effectiveness in filtering out

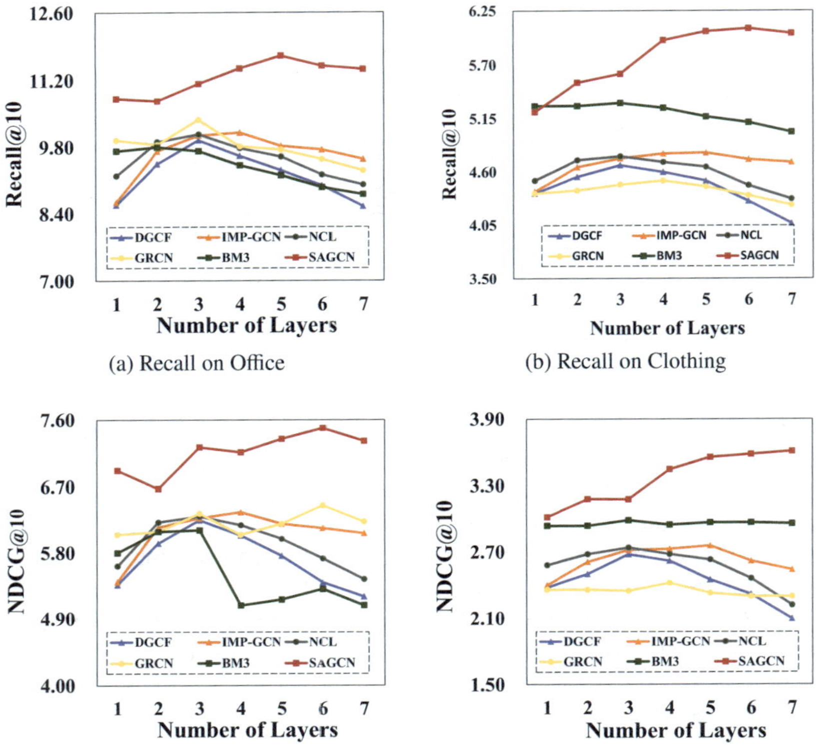

(a) Recall on Office (b) Recall on Clothing

(c) NDCG on Office (d) NDCG on Clothing

Fig. 4.5: Performance Comparison between SAGCN and competitors at different layers on Office and Clothing. Notice that the values are reported by percentage with '%' omitted.

noisy information from high-order neighbors. Our proposed SAGCN consistently outperforms all competitors when more than 2 layers are stacked, and it achieves peak performance when stacking 5 or 6 layers. This indicates that SAGCN can alleviate the over-smoothing problem by performing graph convolution over multiple semantic aspect-based graphs.

4.4.5 Handling of Noisy Neighboring Nodes (RQ4)

We conducted a case study to illustrate how our approach handles noisy neighboring nodes. As shown in Fig. 4.7, we selected user u_0 as the target node and chose some of its first, second, and third-order neighboring nodes to construct a graph displaying their connections. In the original graph, when performing graph convolutions, infor-

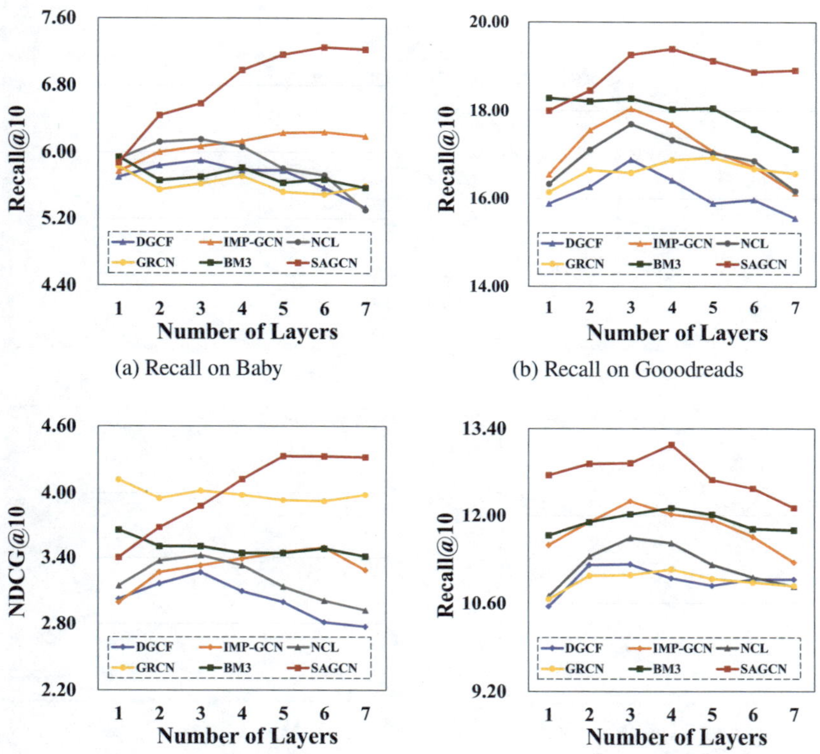

(a) Recall on Baby

(b) Recall on Gooodreads

(c) NDCG on Baby

(d) NDCG on Gooodreads

Fig. 4.6: Performance Comparison between SAGCN and competitors at different layers on Baby and Goodreads. Notice that the values are reported by percentage with '%' omitted.

mation from all neighboring nodes is passed to the target node u_0 through the same graph structure. In contrast, our approach, which leverages LLMs to analyze the semantic aspects mentioned in each review, constructs aspect-aware graphs based on these results. This means that certain neighboring nodes may only appear in specific aspect-aware graphs. For instance, in user u_{2546}'s review of i_{1534}, only the aspect of Price was mentioned, while Ease-of-Use was not. As a result, when constructing the graph, node i_{1534} only appears in the Price aspect-aware graph and passes information to the target node through this specific graph structure.

By constructing aspect-aware graphs based on specific semantic aspects identified in user reviews, our method ensures that only the most relevant information is passed from neighboring nodes to the target node. In traditional graph convolution methods, all neighboring nodes contribute information equally, even if their relevance to the target node is minimal or unrelated, which introduces noise and dilutes the signal from more pertinent nodes. With our approach, neighboring nodes that are not

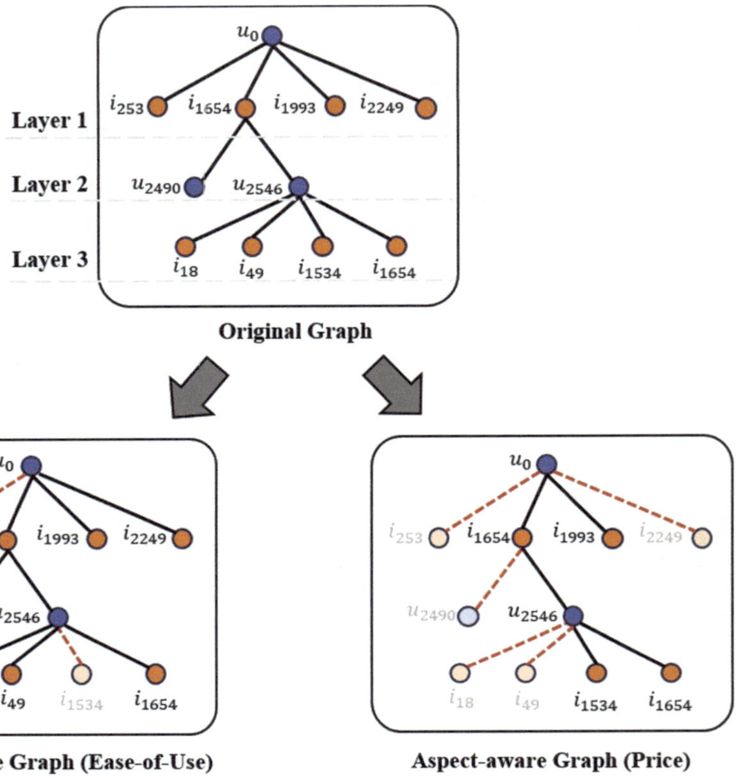

Fig. 4.7: Illustration of Aspect-Aware Graph Construction for Reducing Noisy Neighboring Nodes in Graph Convolutions. User u_0 is the target node, and we selected some of its first, second, and third-order neighboring nodes to show their connectives in the original graph and aspect-aware graphs.

relevant to a specific aspect (*e.g.*, Ease-of-Use or Price) are excluded from the corresponding aspect-aware graph. This selective filtering reduces the contribution of irrelevant or less important nodes, allowing the target node to receive more focused and meaningful information. As a result, the quality of the information passed to the target node improves, leading to more accurate representations and better overall model performance.

4.4.6 Effect of Embedding Dimensions (RQ5)

In this section, we analyze the impact of embedding dimensions on the performance of our proposed method, SAGCN, in comparison with three GCN-based methods (LightGCN, DGCF and DCCF). We experiment with embedding dimensions ranging

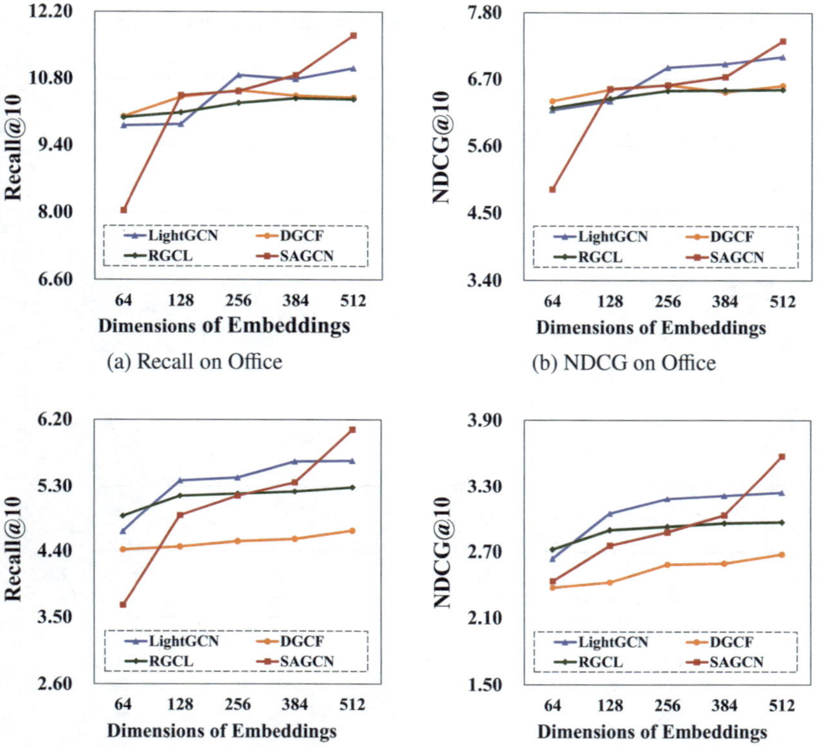

(a) Recall on Office (b) NDCG on Office

(c) Recall on Clothing (d) NDCG on Clothing

Fig. 4.8: Impact of embedding dimensions on the performance of our proposed method, SAGCN, in comparison with three GCN-based methods (LightGCN, DGCF and RGCL). Notice that the values are reported by percentage with '%' omitted.

from 64, 128, 256, 384, 512, evaluating how the size of the embeddings influences model accuracy and overall performance.

From the results shown in Fig. 4.8, we observe a general improvement in performance across all methods as the embedding dimensions increase. However, for LightGCN, DGCF, and RGCL, performance tends to plateau after reaching 128 or 256 dimensions. In contrast, our method, SAGCN, initially performs poorly at 64 dimensions but shows rapid performance growth as the dimensions increase, particularly beyond 384 dimensions. This improvement can be attributed to the aspect-aware graphs constructed using LLMs to analyze user reviews. Our approach transforms simple user interactions into multiple interaction patterns, significantly enriching the informational content of the interaction data, especially after constructing the aspect-aware graphs. As a result, a larger feature space is required to effectively learn user and item representations from these enriched interactions. The increased information not only enhances the performance of our method but also necessitates a larger feature space to capture the complexity of these interactions.

4.4.7 Semantic Aspects Analysis (RQ6)

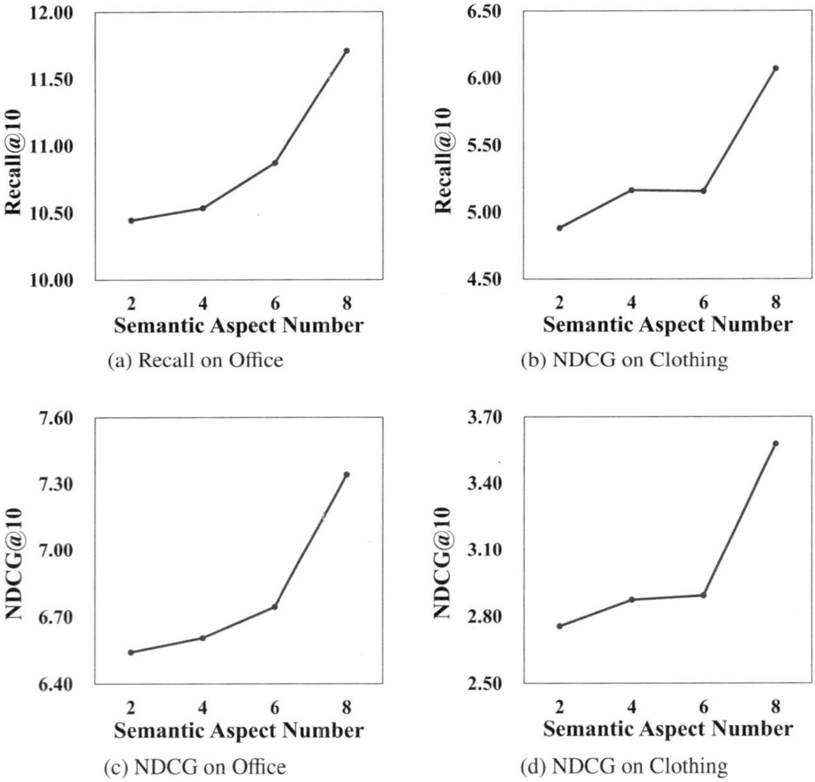

Fig. 4.9: Performance comparison of SAGCN with different semantic aspect numbers on Office and Clothing. Notice that the values are reported by percentage with '%' omitted.

Semantic Aspect-Aware Interactions Analysis

In our experiment, we selected eight aspects per dataset by merging similar aspects and eliminating those with very low occurrence. Table 4.7 reported the number of interactions for different semantic aspects across four datasets.

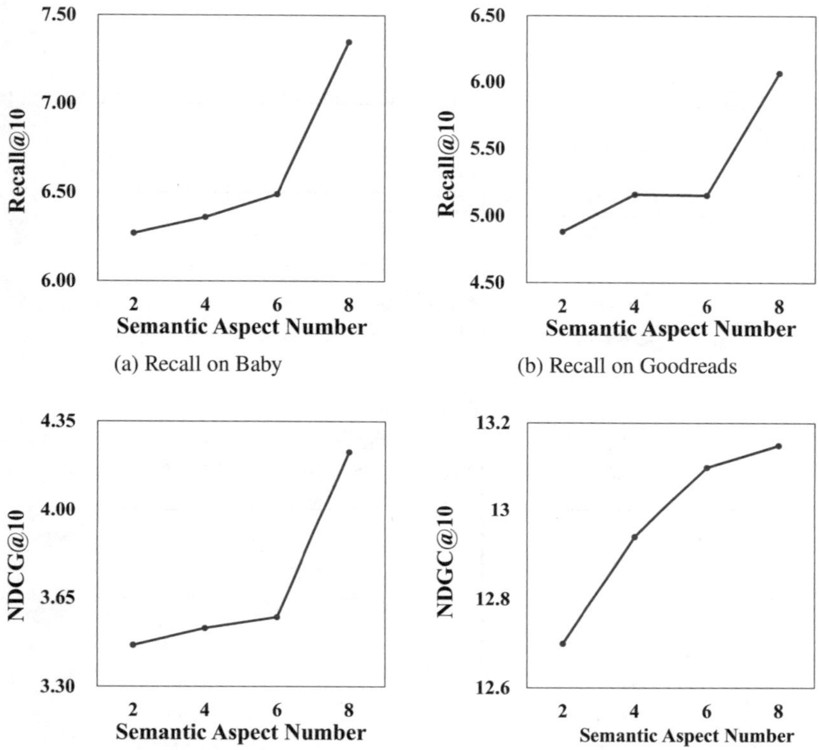

(a) Recall on Baby (b) Recall on Goodreads
(c) NDCG on Baby (d) NDCG on Goodreads

Fig. 4.10: Performance comparison of SAGCN with different semantic aspect numbers on Baby and Goodreads. Notice that the values are reported by percentage with '%' omitted.

Based on our observations of various recommendation scenarios, it is evident that users have distinct concerns regarding different aspects. For instance, in reviews of office products, *convenience* is a common concern, while *fit* is only mentioned in clothing reviews. To rank the semantic aspects in each recommendation scenario, we analyzed their interaction numbers. This helps us identify that the importance of different semantic aspects varies depending on the number of interactions. For example, in reviews of Office products, *Durability* was mentioned more frequently than *Design*. On the other hand, in reviews of Baby, *Design* was mentioned more frequently than *Durability*. Note that the most frequently mentioned aspect being *Quality* in the Office, Baby, and Clothing datasets. For these types of products, customers often prioritize *Quality* because it directly influences their decision-making. In contrast, for GoodReads, the most mentioned aspect is *Satisfaction*, which aligns with the subjective nature of book reviews. Readers tend to express overall satisfaction with the reading experience, which encompasses personal enjoyment, engagement with the story, or alignment with expectations.

Table 4.7: The extracted semantic-aware aspects and the number of their associated interactions over four datasets.

Semantic-aware Aspects	Office	Baby	Clothing	Goodreads
Quality	43,850	133,337	230,162	11,394
Functionality	43,269	132,827	-	-
Ease of Use	42,347	125,014	-	-
Convenience	41,238	-	-	-
Comfort	40,795	127,938	210,254	-
Durability	37,564	116,212	-	-
Design	23,973	119,013	181,222	-
Price	23,661	59,859	106,416	14,903
Appearance	-	-	205,532	19,707
Style	-	-	188,378	-
Fit	-	-	186,132	-
Size	-	86,125	170,869	-
Satisfaction	-	-	-	34,774
Relevance	-	-	-	31,686
Support	-	-	-	19,366
Usefulness	-	-	-	14,111
Creativity	-	-	-	10,254

Effect of Aspect Numbers

To investigate the effect of semantic aspect numbers, we conducted experiments across all four datasets. Note that all aspects were ranked according to the number of interactions, we selected a number of aspects based on their ranking. Our findings, shown in Figure 4.9 and 4.10, suggest that increasing the semantic aspect numbers can improve the performance of SAGCN. In other words, SAGCN achieves its peak performance when leveraging all aspects, indicating that it can benefit from various aspects. Furthermore, we found that SAGCN with eight aspects showed a greater improvement in performance than it with six aspects. This could be due to the fact that the added two aspects might provide more supplementary information to the other six. When combined, the full set of features gives the model a richer understanding of user behaviors.

4.4.8 Interpretability Analysis (RQ7)

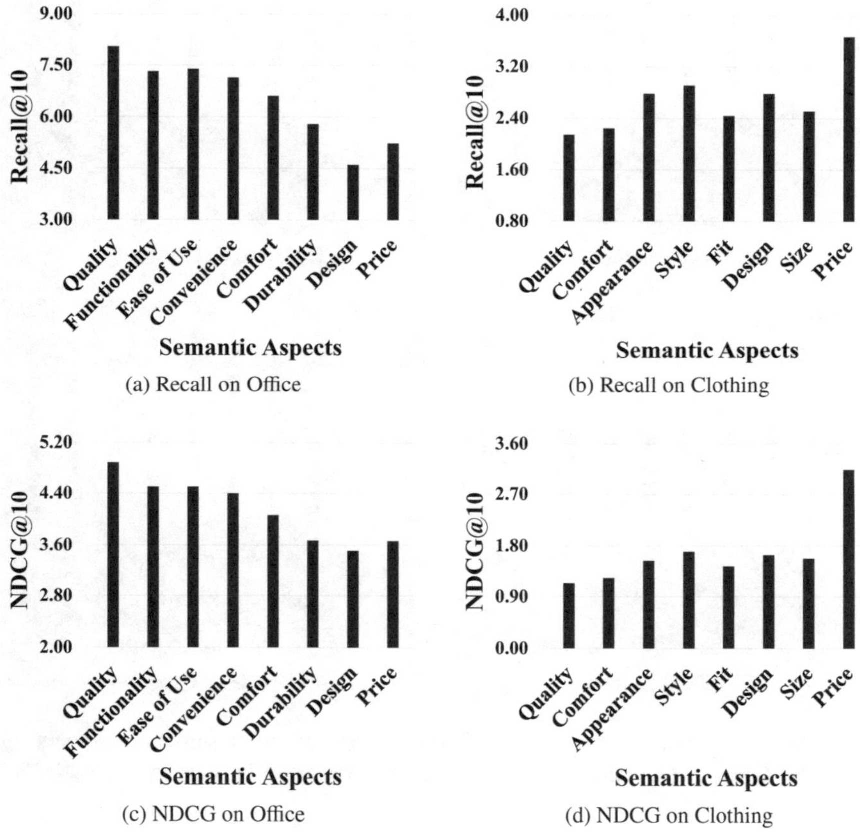

(a) Recall on Office

(b) Recall on Clothing

(c) NDCG on Office

(d) NDCG on Clothing

Fig. 4.11: Performance comparison of SAGCN on different semantic aspects on Office and Clothing. Notice that the values are reported by percentage with '%' omitted.

In this section, we examined the interpretability of our model from two different perspectives. First, our model can interpret the contribution of each semantic aspect in a recommendation scenario. To examine the contribution of different aspects, we predicted the users' preferences using individual aspect embeddings of the users and items. Fig. 4.11 and 4.10 show the following interesting observations: (1) Different semantic aspect features have varying contributions in recommendation. For instance, *Quality* has a greater impact than *Design* for Office. (2) The number of semantic-aware interactions is not the main contribution factor. Surprisingly, even though the number of interactions for *Price* is the least, it still has the highest

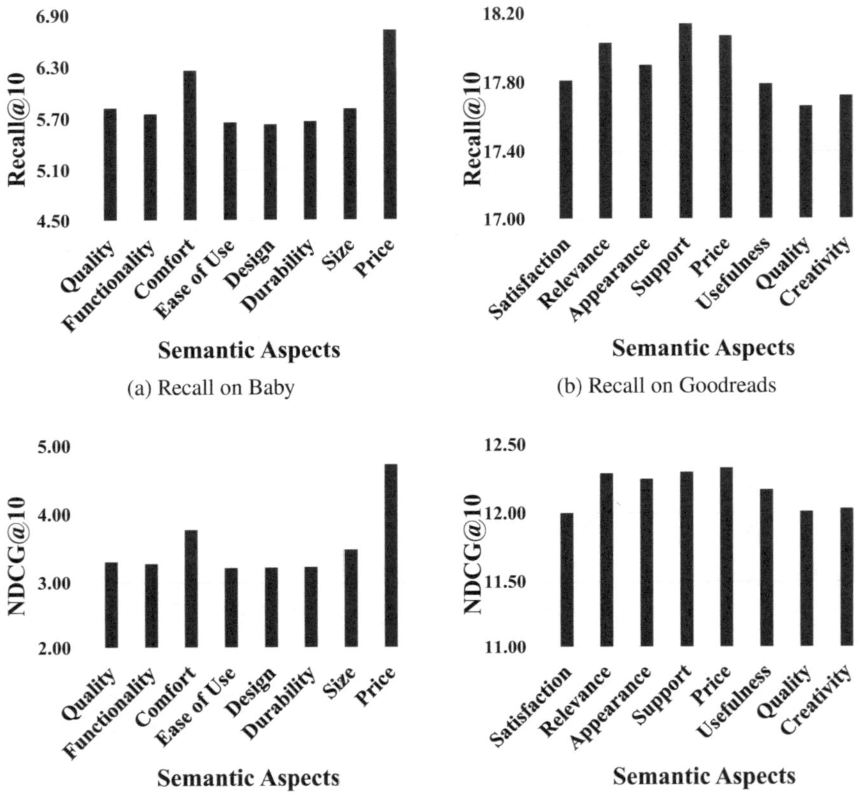

(a) Recall on Baby
(b) Recall on Goodreads
(c) NDCG on Baby
(d) NDCG on Goodreads

Fig. 4.12: Performance comparison of SAGCN on different semantic aspects on Baby and Goodreads. Notice that the values are reported by percentage with '%' omitted.

contribution to our model. This could be because *Price* introduces more valuable interactions than the other aspects.

We can further provide a detailed explanation by providing the scores of various semantic aspects for each user-item pair. In Fig. 4.13 (a), we computed the preference score of user $u3547$ for items $i836$ and $i1322$ from different semantic aspects in Office. From the result, we can conclude that the user has similar but different preferences on different aspects of different items. Moreover, the user would be interested in these two items mainly due to their *Convenience* and *Quality*. Additionally, the aspect of *Comfort* in $i1322$ might be more appealing to the user than $i836$. Similarly, we computed the preference score of users $u4842$ and $u2071$ for item $i836$, which is illustrated in Fig. 4.13 (b). These two users have diverse preferences for the same item. Specifically, user $u4842$ focuses on *Design*, *Price* and *Convenience*. In

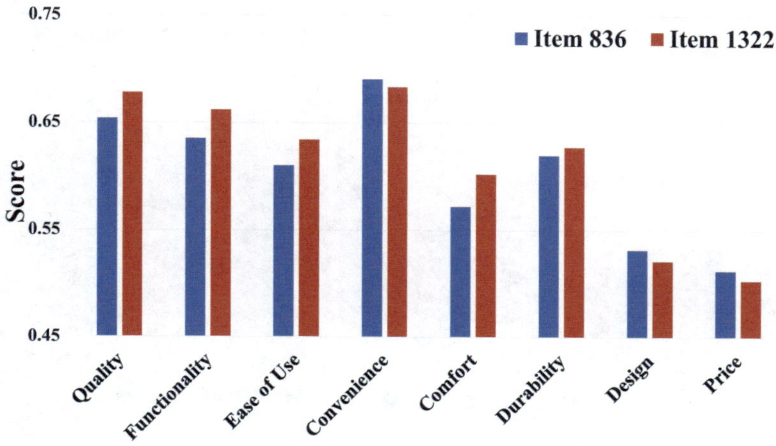

(a) User (u3547)

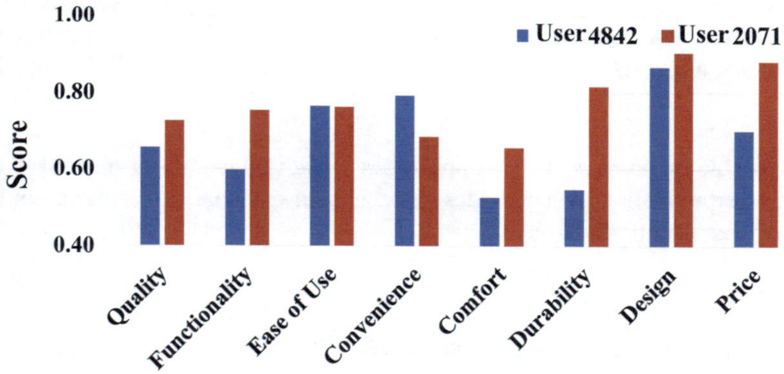

(b) Item (i2210)

Fig. 4.13: The examples of preference scores across different semantic aspects in Office. a) Preferences of user $u3547$ for item $i836$ and $i1322$ on different semantic aspects. b) Preferences of user $u4842$ and $u2071$ for item $i2210$ on different semantic aspects. Notice that the values are reported by percentage with '%' omitted.

contrast, user $u2071$ pays more attention to the *Design*, *Price*, and *Durability*. These experimental results also demonstrate the interpretability of our proposed approach.

Overall, we further emphasize our contribution to interpretability compared to traditional methods using disentanglement techniques. As shown in Fig. 4.14, in disentanglement-based methods, each representation chunk is associated with a latent factor that cannot be understood by humans. In other words, while we can identify which factor contributes more to user preferences, we have no understanding of what that factor actually represents. In contrast, our approach associates the representations learned from different graphs with specific semantic aspects, which are directly

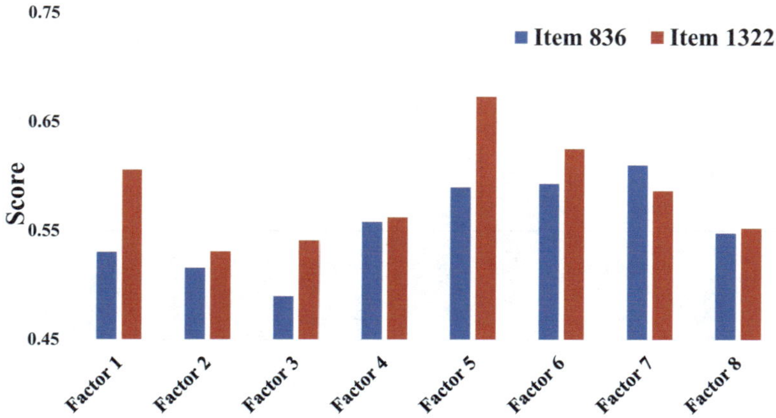

Fig. 4.14: The example of preference scores across different aspects associated with latent factors in Office. Preferences of user $u3547$ for item $i836$ and $i1322$ on different latent factors. Notice that the values are reported by percentage with '%' omitted.

understandable by humans. This demonstrates the advantage of our method in enhancing interpretability, as it provides clear, human-readable explanations for the factors driving user preferences.

4.4.9 Feature Independence Analysis (RQ8)

We present a visualization of the feature independence of different semantic aspects in Office and Clothing, as shown in Fig. 4.13. To achieve this, we first sampled one user from the user set and then calculated the cosine similarity scores among different aspects based on the user's representation vector. The heatmap in Fig. 4.13 shows that the similarity values are close to zero between different aspects. This implies that the representations associated with different semantic aspects are independent. This demonstrates that the SAGCN model can learn robust representations with the semantic aspects extracted via LLM, enabling more accurate item recommendations for users. An additional noteworthy finding is that the representations are unaffected by the potential bias stemming from the number of semantic aspect-aware interactions. For example, as indicated in Tab. 4.7, there is an imbalance in the number of aspect-aware interactions(*e.g.*, the *Quality* aspect-aware interactions occupy a large proportion in all aspect-aware interactions on both Office and Clothing). In previous data-driven recommendation models, such an imbalance might lead the model to overly focus on these dominant aspects. However, the SAGCN approach leverages the aspect-aware interactions solely for constructing the semantic-aware graph,

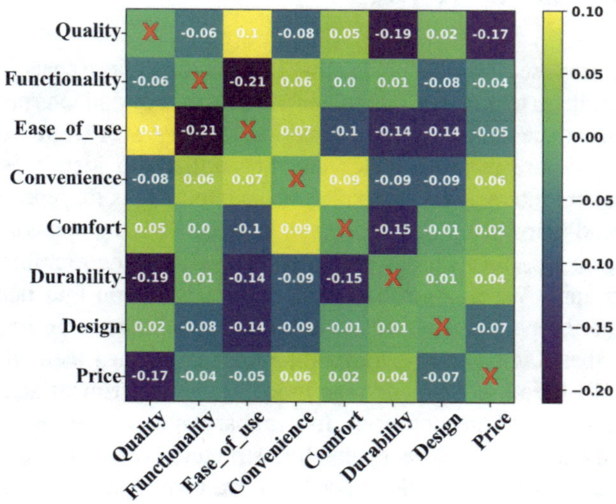

(a) Similarity scores among different aspects (Office)

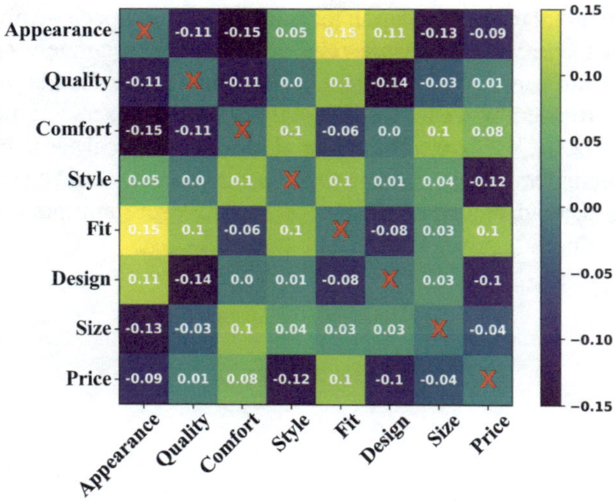

(b) Similarity scores among different aspects (Clothing)

Fig. 4.15: Visualization of the feature independence of semantic aspects.

meaning that the quantity of these interactions does not sway the aspect-associated representations. This further demonstrates that our proposed model can obtain more robust and better representations by capturing the fine-grain relations between each user and item at the semantic aspect level.

4.5 Conclusion and Future Work

In this chapter, we propose a novel approach that extracts semantic aspects from user reviews and utilizes them to enhance recommendation accuracy and interpretability. Specifically, we introduce a chain-based prompting strategy with LLMs to extract the semantic aspect-aware reviews, which are then utilized to identify semantic aspect-aware user-item interactions. To exploit these interactions for representation learning of users and items, we propose a semantic aspect-based graph convolution network (SAGCN) that performs graph convolution operations on multiple semantic aspect-based graphs. We conducted extensive experiments on four benchmark datasets to evaluate the performance of our SAGCN model. Our experimental results demonstrate that SAGCN outperforms the pure LLM-based method and the state-of-the-art GCN-based recommendation models. The experiment analysis exhibits the interpretability of our approach. In summary, our work emphasizes the importance of analyzing user reviews by creating structured reviews to understand users before recommendations. On the other hand, we investigate a new approach to improving recommendation performance by integrating the capability of LLMs with conventional recommendation techniques.

This work is an initial attempt to combine the sentiment analysis capabilities of LLMs with conventional recommendation techniques. Our approach involves using a frozen LLM to identify aspect-aware reviews. However, accurately extracting semantic aspects from user reviews remains a challenging task. In addition, apart from the reviews provided by users, the interactions between users and items also include semantic aspects. We expect the potential of sentiment analysis to be further explored for a better understanding of user intent. Furthermore, it would be interesting to explore how to take advantage of these aspect-aware user-item interactions for representation learning.

Chapter 5
Attribute-aware Attentive Graph Convolutional Network

5.1 Introduction

Beyond their ability to learn robust user and item representations, Graph Convolutional Networks (GCNs) have shown remarkable potential in solving diverse challenges faced by recommender systems. One particularly pressing challenge is the attribute missing problem, where incomplete or unavailable user or item attributes can hinder the accuracy and effectiveness of recommendations. In this chapter, we delve into how GCNs can be utilized to address this issue by effectively propagating and aggregating information across graph structures.

The performance of Collaborative Filtering (CF)-based recommendation methods is often unsatisfactory when the interactions are sparse. To tackle this issue, an effective solution is to leverage the related side information (such as reviews, images, social relations, and attributes), which provides additionally rich information for user preference and item feature modeling [111]. Attributes are the most exploited among various side information, including the category and brand of products, genre and directors of movies, as well as descriptive tags of social images. Many methods have been developed to leverage above attributes in recommendation [111, 19, 53, 117], because they provide important and concise information for items. In general, existing methods are mainly in two paradigms according to the ways of exploiting attributes. One is to convert attributes into binary feature vectors via one-hot/multi-hot encoding based on generalized linear models [111, 19], such as logistic regression, support vector machines, linear part of factorization machines (FMs)[111], and Wide&Deep [19]. Despite their success, their defect is that the features become very sparse because of the one-hot/multi-hot encoding, making it difficult to learn reliable parameters [111]. Besides, they also significantly increase the feature space. For instance, in Amazon's Kindle Store dataset, there are 1,438 attribute labels in total. A 1,438-dimension feature vector will be generated via the multi-hot embedding method and only few dimensions are non-zeros in this feature vector. Another way is to first embed each attribute into a k-dimension feature vector, and then concatenate the feature vectors of all attributes as an input feature vector

F. Liu, L. Nie, *Advancing Recommender Systems with Graph Convolutional Networks*,
https://doi.org/10.1007/978-3-031-85093-6_5

for subsequent modules [53, 117, 19]. For example, NFM [53] projects each feature to a dense representation via the embedding layer. This approach is widely-adopted in recently proposed deep learning based models[117, 19], which can alleviate the aforementioned sparsity problem. However, some shortcomings still exist with the above two kinds of methods.

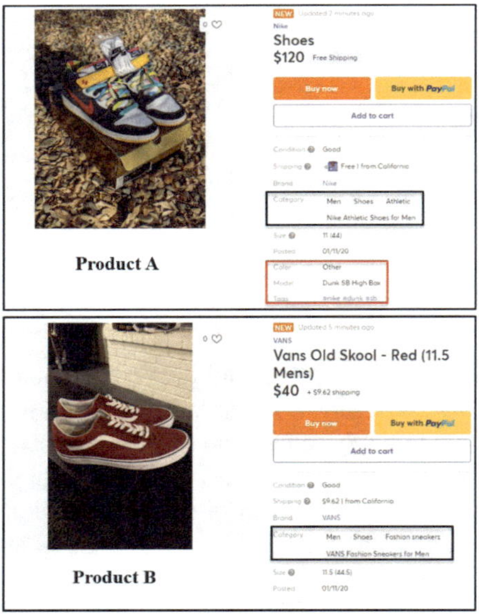

Fig. 5.1: Two items with different attribute information from Mercari.

Firstly, these methods cannot well tackle the attribute missing problem. It is well established that the provided attributes of items are incomplete in real systems. Fig. 5.1 shows the attributes of two products from Mercari [1], a second-hand product E-commerce platform. As we can see, each product misses some attributes. For example, Product B misses the *Color*, *Model* and *Tags*. The common solutions to deal with this problem in existing models [111, 53, 19, 117] are: 1) *substituting the missing attribute with a default value* [111, 53, 19] like "Other", as demonstrated in Fig. 5.2. Considering that the substituted attribute is meaningless and the default value will be used for different items with different attributes, the replaced attribute cannot properly describe item and will mislead the embedding learning. And 2) *simply assuming that the item does not possess the attribute* [53, 19, 117], as illustrated by the examples in Table 5.1. Obviously, this assumption introduces misleading information to the model. In general, the strategies of the existing models on dealing with missing attributes are inappropriate and may inject biased information to the embedding learning, resulting in sub-optimal performance.

[1] https://www.mercari.com

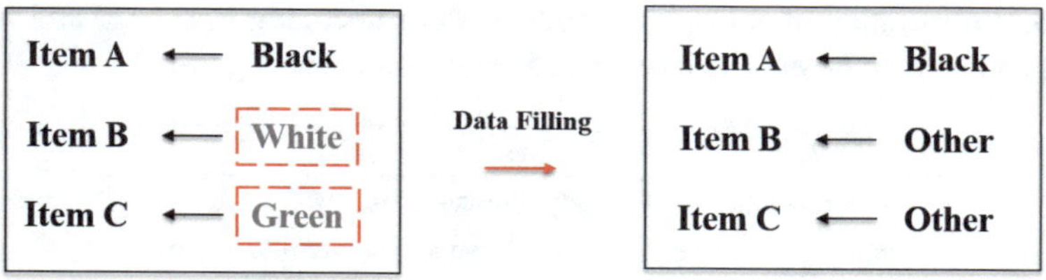

Fig. 5.2: A toy example of substituting the missing attributes with a default value. Red rectangle indicates that the attribute is missing.

Secondly, for most of previous methods, they treat all attributes equally for both items and users. In those methods, such as NFM [53] and Wide&Deep [19], the embeddings of all attributes are concatenated and projected into a unified feature space without differentiating their importance. We deem that some attributes are more important for an item. What is more, different users may attach importance to different attributes of an item. For example, one may pay more attention to the appearance of a product, such as color and style, while another may care more about the material or price.

Motivated by the above analyses, in this chapter, we present an Attribute-aware Attentive Graph Convolution Network (A^2-GCN for short), which seamlessly incorporates the item attribute information into recommendation[2]. To be more specific, in our model, the users, items and attributes are treated as three types of nodes to construct the graph. Users and items are linked based on their interactions, and attributes are linked to their associated items. In addition, the message-passing strategy [5] is used to learn the node representation by aggregating information passed from its neighbor nodes. As to the user representation, it is learned by aggregating messages from neighbor items. Similarly, the item representation is learned by aggregating messages from its neighbor users and attributes. Considering a user may have different preferences for different items, we introduce an attention method to filter the messages passed from different neighbor nodes. Furthermore, the attributes of an item can affect a user's preference for this item. To model this effect, when computing the attention of an item to a user, we design an *attribute-aware attention mechanism* to characterize the attribute information of the given item. As our model only leverages the available attribute information (based on the item-attribute connections in the graph) to learn user and item representations, the first aforementioned problem can be naturally avoided. Besides, we devise an attention mechanism to estimate the importance of different attributes, which addresses the second problem in previous methods. Extensive experiments on several large-scale and real-world datasets have been conducted to demonstrate the effectiveness of our model by comparing it with several strong baselines.

In summary, the main contributions of this work are as follows:

---

[2] Our model can be easily extended to consider user attribute information, such as gender and age. In this work, we focus on item attributes, since they are easy to access and more widely used in recommender systems.

Table 5.1: Illustration of existing methods on the embedding with (E_w) and without ($\mathrm{E}_{w/o}$) an attribute label. The label in red denotes the missing one.

Movies	Tags	E_w	$\mathrm{E}_{w/o}$
Grumpier Old Men	Comedy; Romance	$v_c v_r \overline{v}_d$	$v_c v_r \overline{v}_d$
Waiting to Exhale	Comedy; Drama	$\boldsymbol{v_c \overline{v}_r v_d}$	$\boldsymbol{\overline{v}_c \overline{v}_r v_d}$
Nixon	Drama	$\overline{v}_c \overline{v}_r v_d$	$\overline{v}_c \overline{v}_r v_d$

- We step into the existing attribute-aware recommendation models and analyze their shortcomings. Inspired by that, we present a new GCN-based model called A^2-GCN, which can naturally address the attribute missing problem in real-world datasets.
- We highlight the importance of attribute information on user preference and design a novel attribute-aware attention mechanism to capture this effect.
- We have conducted extensive experiments on four real-world datasets to demonstrate the superiority of our model over several state-of-the-art baselines.

The rest of this chapter is structured as follows. In Section 5.2, we briefly review the related literature. Section 5.3 introduces the preliminaries relevant to our proposed model. In Section 5.4, we elaborate the proposed model followed by experiments in Section 5.5. We finally conclude this chapter in Section 5.6.

5.2 Related Work

In this section, we briefly reviewed the recent advancement of model-based collaborative filtering model, especially the attribute-aware and GCN-based methods for specific recommendation scenarios, which are most close to our work.

Model-based collaborative filtering methods [56, 148, 154, 55, 43, 21, 61, 78] learn user and item representations based on the user-item interactions for recommendation. This paradigm has achieved great success since the matrix factorization [75] stood out in the Netflix prize contest [4]. Although these methods have greatly enhanced the performance, they still suffer from the sparsity problem because they merely rely on the interaction data. A common solution is to leverage side information, such as reviews and attributes, to assist the user and item representation learning, because side information can provide additionally valuable information. In the next, we mainly discussed the attribute-aware and GCN-based methods for specific recommendation scenarios.

5.2.1 Attribute-aware Recommendation Models

Attributes are widely available and valuable for item description. Thus, many methods have been developed to leverage attributes in recommendation [111, 19, 47, 165]. Factorization Machines (FMs) based models can model the second-order interactions among attributes via the inner product of their factorized vectors [111, 107, 53, 152, 20]. Moreover, they can work with any real-valued feature vectors by modelling all interactions within each pair of features via factorized interaction parameters for prediction. As the inner product violates the triangle inequality and cannot capture finer-grained attribute interactions, TransFM [107] substitutes the squared Euclidean distance function for the inner product for sequential recommendation. The performance of Factorization Machine (FM) [111] is limited by its linearity and the modelling of pairwise feature interactions. To deal with the linearity of FM, NFM [53] is proposed to handle the complex and underlying nonlinear structure in real-world dataset. In this method, the attribute information, such as user and item features, is used as side information and embedded into different vectors. AFM [152] extends NFM by adding an attention mechanism to discriminate the importance of different feature interactions.

Another line of research adopts the deep neural network (DNN) techniques to learn the attribute embeddings and concatenate them for recommendation [19, 116, 117, 15]. In particular, Wide&Deep [19] combines the deep neural network and the linear model for recommendation, where the deep part is a multi-layer perceptron (MLP) on the concatenation of feature embedding vectors to learn feature interactions. DeepCross [116] shares a similar framework with Wide&Deep by replacing the MLP with the state-of-the-art residual network. ACCM[117] adaptively adjusts the source information of each user-item pair with an attention mechanism.

In brief, the above methods suffer from the two limitations discussed in Section 1. In this chapter, we developed a GCN-based attribute-aware recommendation model, which can effectively tackle those limitations.

5.2.2 GCN-based Models for Specific Recommendation Scenarios

In recent years, Graph Convolutional Networks (GCNs) have achieved great success due to the powerful capability on representation learning from non-Euclidean structure [157, 171, 38, 5, 135, 137, 36, 138, 170, 77, 155, 23]. There are also GCN-based models specially designed for specific recommendation scenarios, such as social recommendation [38], micro-video recommendation [140], and knowledge-aware recommendation [135]. In particular, GraphRec [38] introduces a method to consider heterogeneous strengths of social relations for social recommendation. GCN-PHR [140] leverages GCN techniques to model the complicated interactions among <users, hashtags, micro-videos> and learn their representations. KGAT [135] models the high-order connectivities in Knowledge Graph based on an end-to-end learning framework. In addition, HAN [137] aggregates features from meta-path based neighbors in a hierarchical manner to generate node embedding.

In this chapter, we mainly focused on the attribute-missing problem in attribute-aware recommendation and presented an attribute-aware attentive GCN recommendation model. Our model can leverage attribute information of items to learn better user and item representation for improving recommendation performance. Especially, we design an attention mechanism to estimate the influence of neighbor nodes on items, and an attribute-aware attention mechanism to incorporate the attribute embeddings to compute the influence of neighbor item nodes on users. Besides, our model can be applied to different scenarios with attribute information, such as E-commerce product, social images, and videos.

5.3 Preliminaries

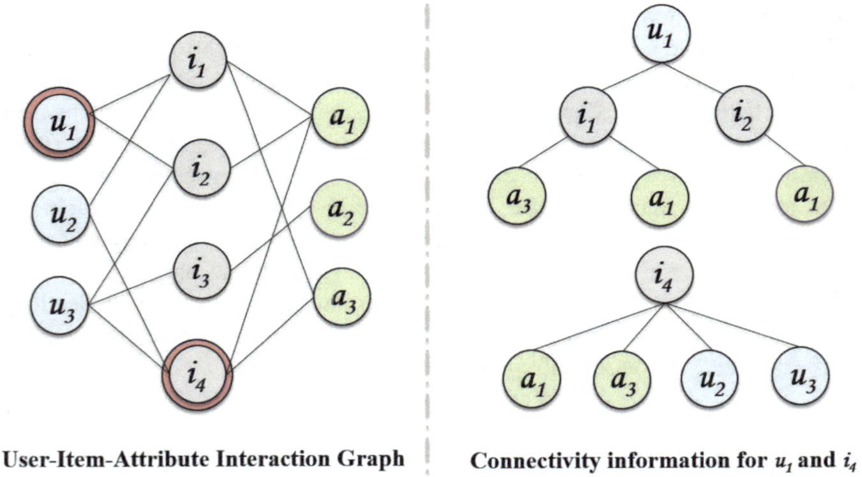

Fig. 5.3: A toy example of an interaction graph based on <users, items, attributes> and connectivity information for u_1 and i_4. Blue, grey and green circles denote users, items and attributes, respectively. The node u_1 and node i_4 are respectively the target user and item in this example.

Before describing our model, we would like to introduce the problem setting first. A dataset with an interaction matrix $\boldsymbol{R}^{N_u \times N_v}$ of a set of users $\mathcal{U}$ and a set of items $\mathcal{V}$ is involved, where N_u and N_v are the numbers of users and items, respectively. In the dataset, each item is associated with a set of attribute labels $a \in \mathcal{A}$ which describe different attributes of the item, where the number of attribute labels is defined as N_a. The matrix $\boldsymbol{R}$ records the interaction history between users and items. A nonzero entry $r_{uv} \in \boldsymbol{R}$ indicates that user $u \in \mathcal{U}$ has interacted with item $v \in \mathcal{V}$ before; otherwise, the entry is zero. Notice that the interactions can be implicit (*e.g.*, click) or explicit (*e.g.*, rating). Based on this dataset, we aim to recommend a user $u \in \mathcal{U}$ with suitable items which the user did not consume before and will be appealing to.

Let $\mathcal{G} = (\mathcal{W}, \mathcal{E})$ be an undirected graph, where $\mathcal{W}$ denotes the set of nodes and $\mathcal{E}$ is the set of edges. Specifically, $\mathcal{W}$ consists of three types of nodes: users $u_i \in \mathcal{U}$ with $i \in \{1, \cdots, N_u\}$, items $v_j \in \mathcal{V}$ with $j \in \{1, \cdots, N_v\}$, attributes $a_k \in \mathcal{A}$ with $k \in \{1, \cdots, N_a\}$, and $\mathcal{U} \cup \mathcal{V} \cup \mathcal{A} = \mathcal{W}$. For the ease of presentation, i, j and k will be assigned to index user, item and attribute, respectively. In the graph, user nodes and item nodes are linked based on their interaction history; item nodes and attribute nodes are linked based on the association of attributes to the corresponding items. Note that there is no edge between user nodes and attribute nodes. Fig. 5.3 illustrates a toy example of an interaction graph which is based on <users, items, attributes>, and the connectivity information for user u_1 and item i_4. We take user u_1 and item i_4 as examples for illustration. User u_1 has neighbor nodes i_1 and i_2; i_1 has attributes a_1, a_3 and i_2 links to its attribute node a_1. For item i_4, it has interactions with user u_2, u_3, as well as attributes a_1 and a_3.

5.4 A^2-GCN Model

5.4.1 Model Overview

Typically, model-based CF methods learn the representations of users and items by collectively exploiting the user-item interactions. To be more specific, the attribute provides valuable information about item features, which can be leveraged to capture user preferences. The attribute information of an item can affect a user's preference for the item, because different users may prefer different aspects of an item and attributes deliver item information from different aspects. For example, some users like a *movie* because of its *director* while others because of the *leading actor*. Therefore, different users focus on different features of the movie because of its attribute information, or in other words, the attributes of *director* and *actor* pass different information of the movie to different users. Therefore, to accurately model user and item representations, a desired model should capture the complex interactions among users, items, and attributes.

In this chapter, we refer to the graph convolutional network (GCN) techniques, because of its powerful representation learning capability, to learn user and item representation in the graph. The core idea behind the node representation learning in GCN is how to recursively aggregate feature information from local graph neighborhoods via neural networks. In our cases, an item node collects information from the connected user nodes and attribute nodes to learn its representation. In particular, given that different users and attributes provide information from different aspects and contribute to an item with different importance, an attention mechanism is designed to distill useful information passed from user nodes and attribute nodes to the item. Similarly, the representation of a user node is learned by distilling information from the connected item nodes which represent the ones this user has consumed before. As discussed above, the attributes of an item can have important effects on a user's preference for this item. To model the effects, we propose a novel attribute-aware attention mechanism which leverages the information of the attribute nodes connected to the items to help learn the attention vector of this item with respect to the target user. Towards this end, our model can well exploit the interactions among users, items, and attributes for user and item representation learning.

5.4.2 Item Representation Learning

We adopt the messaging-passing strategy [5] to learn user and item representation. Let $\boldsymbol{e_u} \in \mathbb{R}^d$, $\boldsymbol{e_v} \in \mathbb{R}^d$, and $\boldsymbol{e_a} \in \mathbb{R}^d$ be the embeddings of user u, item v, and attribute a, respectively. d is the embedding size. In the next, we detail the algorithm for user and item representation learning[3].

For an item v, the representation is learned by aggregating all the information passed from the connected users and attributes. The learning process of item representation is described in the following.

Message passing

In our model, the representation of an item v is modeled by accumulating the incoming messages from all the user neighbors $u \in \mathcal{N}_u^v$ and attribute neighbors $a \in \mathcal{N}_a^v$, where $\mathcal{N}_u^v$ and $\mathcal{N}_a^v$ are the user neighbor set and attribute neighbor set of item v, respectively. Based on the idea of message passing, the information passed from a user neighbor $u \in \mathcal{N}_u^v$ is defined as:

$$\boldsymbol{m_{v \leftarrow u}} = \gamma_u^v(\boldsymbol{W_1 e_u} + \boldsymbol{W_2}\,(\boldsymbol{e_u} \odot \boldsymbol{e_v})), \tag{5.1}$$

where $\boldsymbol{m_{v \leftarrow u}}$ is the embedding vector of the message passed from user u to item v. W_1 and $W_2 \in R^{d' \times d}$ are trainable weight matrices which can distill useful message from $\boldsymbol{u_i}$. d' is the transformation size. In the message-passing process, we consider the interaction between two nodes via $\boldsymbol{e_u} \odot \boldsymbol{e_v}$, where $\odot$ denotes the element-wise product. It makes the message dependent on the affinity between u_i and v_j, *e.g.*, passing more messages from the similar items. This strategy can increase the model representation ability and thus boost the performance for recommendation as demonstrated in [136]. γ_u^v is a parameter to control how much information to be passed from this user to the item, which is computed by an attention mechanism that will be introduced later.

---

[3] Given that the primary goal of this work is to study the effectiveness of our model on exploiting attribute information for user and representation learning, our model does not learn the attribute embeddings. However, the embedding of attribute nodes can also be updated in a similar way as user or item nodes do.

In the same way, the message passed from attribute node a to item node v is obtained by

$$\boldsymbol{m}_{\boldsymbol{v}\leftarrow\boldsymbol{a}} = \gamma_a^v(\boldsymbol{W}_{\boldsymbol{1}}\boldsymbol{e}_{\boldsymbol{a}} + \boldsymbol{W}_{\boldsymbol{2}}\left(\boldsymbol{e}_{\boldsymbol{a}} \odot \boldsymbol{e}_{\boldsymbol{v}}\right)). \tag{5.2}$$

The notations in this equation are defined in the same manner as in the Eq. 5.1.

Besides, the original information of the node is also important. To retain the information of $\boldsymbol{v}$, we add a self-connection to node v to recycle the information, which is formulated as follows:

$$\boldsymbol{m}_{\boldsymbol{v}\leftarrow\boldsymbol{v}} = \gamma_v^v \boldsymbol{W}_{\boldsymbol{1}}\boldsymbol{e}_{\boldsymbol{v}}. \tag{5.3}$$

Message aggregation

Aggregating all messages passed from user u and attribute a, item v's embedding $\boldsymbol{e}_{\boldsymbol{v}}$ is updated by

$$\boldsymbol{e}_{\boldsymbol{v}} = \text{LeakyReLU}\left(\boldsymbol{m}_{\boldsymbol{v}\leftarrow\boldsymbol{v}} + \sum_{u\in\mathcal{N}_u^v} \boldsymbol{m}_{\boldsymbol{v}\leftarrow\boldsymbol{u}} + \sum_{a\in\mathcal{N}_a^v} \boldsymbol{m}_{\boldsymbol{v}\leftarrow\boldsymbol{a}}\right). \tag{5.4}$$

We use the activation function LeakyReLU [100] to encode both positive and small negative signals into messages.

Attention mechanism

We assume that different neighbor nodes (user nodes and attribute nodes) have different influences on item v. Inspired by this assumption, we design an attention mechanism to estimate the influence of neighbor nodes on items. The influence from user u to item v is formulated as follows:

$$s_u^v = g\left(\boldsymbol{e}_{\boldsymbol{v}}, \boldsymbol{W}_{\boldsymbol{u}}^{\boldsymbol{v}}\boldsymbol{e}_{\boldsymbol{u}}\right), \tag{5.5}$$

where $\boldsymbol{W}_{\boldsymbol{u}}^{\boldsymbol{v}}$ is a weight matrix, and $g\left(\cdot\right)$ is a similarity function to measure the similarity of vectors, which represents the relation of passing message from the user node to the item node. In particular, the cosine similarity function is applied in our work.

Similarly, the influence of attribute nodes a on item nodes v can be defined as s_a^v:

$$s_a^v = g\left(\boldsymbol{e}_{\boldsymbol{v}}, \boldsymbol{W}_{\boldsymbol{a}}^{\boldsymbol{v}}\boldsymbol{e}_{\boldsymbol{a}}\right). \tag{5.6}$$

The notations in this equation are defined in the same manner as in Eq. 5.5. Finally, we normalize the weight s_a^v to obtain the contribution of each user to item representation:

$$\gamma_u^v = \frac{\exp\left(s_u^v\right)}{\sum_{v'\in\mathcal{N}_u^v}\exp\left(s_{u'}^v\right) + \sum_{a'\in\mathcal{N}_a^v}\exp\left(s_{a'}^v\right) + s_v^v}, \tag{5.7}$$

where s_v^v is the weight for the self-connection within item v. Similarly, the contribution of each attribute to item representation γ_a^v and the weight of retaining information by self-connection γ_v^v can be obtained in the same way.

5.4.3 User Representation Learning

For a user u, the representation is learned by aggregating all the information passed from the connected items. The learning process of user representation is the same as that of item representation.

Message passing

The message passed from an item v to a user u is

$$\boldsymbol{m}_{\boldsymbol{u} \leftarrow \boldsymbol{v}} = \gamma_v^u \left(\boldsymbol{W}_{\mathbf{1}} \boldsymbol{e}_{\boldsymbol{v}} + \boldsymbol{W}_{\mathbf{2}} \left(\boldsymbol{e}_{\boldsymbol{u}} \odot \boldsymbol{e}_{\boldsymbol{v}} \right) \right), \tag{5.8}$$

where γ_v^u is to control the amount of information passed from item v to user u. It is computed by the designed attribute-aware attention method described in the next.

Message aggregation

The user representation is updated by aggregating all messages passed from the connected items:

$$\boldsymbol{e}_{\boldsymbol{u}} = \text{LeakyReLU} \left(\boldsymbol{m}_{\boldsymbol{u} \leftarrow \boldsymbol{u}} + \sum_{v \in \mathcal{N}_v^u} \boldsymbol{m}_{\boldsymbol{u} \leftarrow \boldsymbol{v}} \right), \tag{5.9}$$

where $\boldsymbol{m}_{\boldsymbol{u} \leftarrow \boldsymbol{u}}$ is the retained message, and $\mathcal{N}_v^u$ is the set of user u's neighbors.

Attribute-aware attention mechanism

Notice that γ_v^u can be computed using the same attention mechanism as in the item embedding learning as,

$$s_v^u = g(\boldsymbol{e}_{\boldsymbol{u}}, \boldsymbol{W}_{\boldsymbol{v}}^{\boldsymbol{u}} \boldsymbol{e}_{\boldsymbol{v}}), \tag{5.10}$$

where $\boldsymbol{W}_{\boldsymbol{v}}^{\boldsymbol{u}}$ is the weight matrix and $g(\cdot)$ is a similarity function to measure the similarity of vectors. And then γ_v^u is obtained by the softmax normalization on s_v^u:

$$\gamma_v^u = \frac{\exp\left(s_v^u\right)}{\sum_{v' \in \mathcal{N}_v^u} \exp\left(s_{v'}^u\right) + s_u^u}, \tag{5.11}$$

where s_u^u is the weight of the self-connection (*i.e.*, $\boldsymbol{m}_{u \leftarrow u}$). As aforementioned, the attributes of an item can affect a user's preference for the item, however, this method cannot capture this effect. From the perspective of message passing, a user's representation is learned based on the messages passed from all the interacted items (*i.e.*, item embedding). Since attributes have important influence on user preference, the attributes of an item should be incorporated to the message distilling process from the item to the target user. Based on this consideration, we design an attribute-aware attention mechanism, which incorporates the attribute embeddings to compute the γ_v^u. Formally, the weight between v and u is estimated by

$$s_v^u = g\left(\boldsymbol{e_u}, \boldsymbol{W_v^u}\left(\boldsymbol{e_v} + \boldsymbol{W_a^v}\boldsymbol{e_a^v}\right)\right), \tag{5.12}$$

where $\boldsymbol{W_v^u}$ and $\boldsymbol{W_a^v}$ are weight matrices, and $g\left(\cdot\right)$ is a cosine similarity function to measure the similarity of vectors. $\boldsymbol{e_a^v}$ is a combination vector obtained based on the embeddings of all the attributes associated with the item v. Different combination methods can be used to account for the effects of all the related attribute nodes. For simplicity, we use a mean-pooling method

$$\boldsymbol{e_a^v} = \frac{\sum_{a \in \mathcal{N}_a^v} \boldsymbol{e_a}}{|\mathcal{N}_a^v|}. \tag{5.13}$$

To this end, the influence of all the attribute nodes associated with the item is considered in our model in learning the user representation.

5.4.4 Discussion

As we can see, the attribute information has been integrated in our A²-GCN model for user and item embedding learning based on their connections to items (and second-order user neighbors). It is worth mentioning that our model can naturally avoid the limitations in existing attribute-aware methods, which are *using a default value to represent the missing attributes* (*e.g.*, "Other") or *simply assuming that the item does not possess the properties of the missing attributes* (*e.g.*, "Comedy" for the movie "Waiting to Exhale") as discussed in Section 1. In our model, we do not make any assumption on the missing attributes and only consider the available attributes, because only the linked attribute nodes are directly leveraged for the embedding learning. In this way, we can avoid introducing misleading information to the model by inappropriate assumptions, which may hurt the performance. Besides, our model enables the items with missing attributes to benefit from the attribute information propagated from other items via the graph connections. Our experiments validate the superiority of our model over other attribute-aware recommendation models on dealing with the attribute missing problem (see Section 5.5.3).

5.4.5 Model Learning

Prediction

After learning the user and item representations, given a user u and a target item v, the user preference for this item is predicted by:

$$\hat{r}_{uv} = \boldsymbol{e}_{\boldsymbol{u}}^{T} \boldsymbol{e}_{\boldsymbol{v}}. \tag{5.14}$$

Objective function

we target at the top-n recommendation, namely, to recommend a set of n top-ranked items which match the target user's preferences. Similar to other rank-oriented recommendation work [168, 136], we adopt the pairwise-based learning method for optimization. Given two user-item pairs: a positive pair (u, v^+) and a negative pair (u, v^-), the former one indicates that user u has consumed item v^+ before, while the latter one means that no interaction exists between u and v^-. The assumption of pairwise learning is that the user u should prefer v^+ to v^-, namely, $\hat{r}_{uv^+} > \hat{r}_{uv^-}$.

The objective function is formulated as:

$$\arg\min \sum_{(\mathbf{u},\mathbf{v}^+,\mathbf{v}^-)\in\mathcal{O}} -\ln \phi(\hat{r}_{uv^+} - \hat{r}_{uv^-}) + \lambda \left\|\Theta\right\|_2^2, \tag{5.15}$$

where $\mathcal{O} = \{(u, v^+, v^-)|(u, v^+) \in \mathcal{R}^+, (u, v^-) \in \mathcal{R}^-\}$ denotes the training set, $\mathcal{R}^+$ indicates the observed interactions between user u and v^+ in the training dataset, and $\mathcal{R}^-$ is the sampled unobserved interaction set. λ and Θ represent the regularization weight and the parameters of the model, respectively. ϕ is the sigmoid function. And the L_2 regularization is used to prevent overfitting.

Model training

To prevent the overfitting, we adopt *message dropout* and *node dropout* in A^2-GCN. In particular, message dropout randomly blocks the messages passed from one node to another node in the training. It makes the model more independent of edges. Similarly, node dropout randomly removes some nodes, and thus all outgoing messages from those nodes are blocked. This technique enables the embeddings more robust against the presence or absence of particular nodes. These two methods have been successfully applied in previous models [5, 136]. The drop ratios of message dropout and node dropout are empirically tuned in practice.

The mini-batch Adam [70] is adopted to optimize the prediction model and update the model parameters. Specifically, for a batch of randomly sampled triplets $(u, v^+, v^-) \in \mathcal{O}$, we first learn the representations of those users and items based on the propagation rules, and then update the model parameters by using the gradients of the loss function.

Matrix-form propagation rule

In order to update the representation for all users and items in an efficient way, we implement the matrix-form of layer-wise propagation rule [109], which can be described as follows:

$$\mathbf{E} = \sigma\left(\left(\mathcal{L} + \mathbf{I}\right)\mathbf{E}\mathbf{W_1} + \mathcal{L}\mathbf{E} \odot \mathbf{E}\mathbf{W_2}\right), \quad (5.16)$$

where $\mathbf{E} \in \mathcal{R}^{(N_u+N_v+N_a)\times d}$ is the representations for users, items and attributes; and $\mathbf{I}$ denotes an identity matrix. $\mathcal{L}$ represents the Laplacian matrix for the <user, item, attribute> based graph, which is formulated as:

$$\mathcal{L} = \begin{bmatrix} \mathbf{0} & \mathbf{Y_{uv}} & \mathbf{0} \\ \mathbf{Y_{vu}} & \mathbf{0} & \mathbf{Y_{va}} \\ \mathbf{0} & \mathbf{Y_{av}} & \mathbf{0} \end{bmatrix}, \quad (5.17)$$

where $\mathbf{Y_{uv}} \in \mathcal{R}^{N_u \times N_v}$ is the attention weight matrix which represents the weights from item nodes to each user node, where $\gamma_v^u \in \mathbf{Y_{uv}}$. Analogously, $\mathbf{Y_{vu}} \in \mathcal{R}^{N_v \times N_u}$, $\mathbf{Y_{va}} \in \mathcal{R}^{N_v \times N_a}$ and $\mathbf{Y_{av}} \in \mathcal{R}^{N_a \times N_v}$ are the attention weight matrices that represent the weights from user nodes to each item node, the weights from attribute nodes to each item node and the weights from item nodes to each attribute node, respectively.

Table 5.2: Basic statistical analysis of the experimental datasets regarding sparsity.

Dataset	#user	#item	#interactions	sparsity
Office Products	4,905	2,420	53,257	98.10%
Clothing	39,387	23,033	278,676	99.97%
Toys Games	19,412	11,924	167,596	99.93%
Kindle Store	68,223	61,934	982,618	99.98%

5.5 Experiments

To validate the effectiveness of our model, we conducted extensive experiments on four public datasets to answer the following research questions:

RQ1: Does our A^2-GCN model outperform state-of-the-art methods on the top-n recommendation task?

RQ2: How does the attribute missing problem affect the performance of our A^2-GCN model?

Table 5.3: Attribute statistics of the experimental datasets. $\#\text{total}_{attr}$, $\#\text{ave.}_{attr}$, $\#\text{max}_{attr}$, and ratio_{attr} represent the total number of unique attributes, the average number of attributes per item, the maximum number of attributes associated with items, and the ratio of items with attributes, respectively.

Dataset	#attribute	$\#\text{total}_{attr}$	$\#\text{ave.}_{attr}$	$\#\text{max}_{attr}$	ratio_{attr}
Office Products	178	8,205	3.39	6	100.00%
Clothing	253	130,997	5.68	16	100.00%
Toys Games	404	26,844	2.25	7	97.11%
Kindle Store	1,438	423,263	6.83	27	99.80%

RQ3: Can our model effectively leverage the attribute information to alleviate the sparsity problem?

RQ4: Can our proposed attribute-aware attention mechanism improve the performance? If can, how much improvement can be achieved?

In the next, we first introduced the experimental setup, and then answered the above research questions in sequence.

5.5.1 Experimental Setup

Datasets

The public Amazon review dataset[4] [103], which has been widely used for recommendation evaluation in previous studies, is adopted for experiments in this work. Four product categories in this dataset are used in our experiments, as shown in Table 5.2 and Table 5.3. To be more specific, we pre-processed the dataset to only keep the items and users with at least 5 interactions. For each observed user-item interaction, we treated it as a positive instance, and then paired it with one negative item which is randomly sampled from items that the user did not consume before. The basic statistics of the four datasets are shown in Table 5.2. As we can see, the datasets are of different sizes and sparsity levels. For example, the *Office Products* dataset is relatively denser than other datasets. The diversity of datasets is useful for analyzing the performance of our method and the competitors in different situations.

In this work, we focused on the top-n recommendation task, which aims to recommend a set of top-n ranked items that will appeal to the target user. For each dataset, we randomly selected 80% of the interactions from each user to construct the training set, and the remaining 20% for testing. From the training set, we randomly selected 10% of interactions as a validation set to tune hyper-parameters. Therefore, we had at least 3 interactions per user for training, and at least 1 interactions per user

[4] http://jmcauley.ucsd.edu/data/amazon.

Table 5.4: Performance of our A^2-GCN model and the competitors over four datasets. The best and the second best results are highlighted in bold.

Datasets	Office Products		Clothing		Toys Games		Kindle Store	
Metrics	HR	NDCG	HR	NDCG	HR	NDCG	HR	NDCG
BPR-MF	0.2218	0.0952	0.0947	0.0549	0.0534	0.0246	0.2987	0.1469
NeuMF	0.2347	0.0997	0.1213	0.0591	0.0578	0.0266	0.3324	0.1607
NGCF	0.2503	0.1077	0.1366	0.0626	0.0610	0.0281	0.3413	0.1659
NFM	0.2446	0.1041	0.1468	0.0643	0.0618	0.0288	0.3506	0.1745
Wide&Deep	0.2512	0.1082	0.1488	0.0656	0.0626	0.0294	0.3588	0.1801
ACCM	0.2523	**0.1092**	0.1438	0.0634	0.0667	0.0313	0.3468	0.1681
ENSFM	**0.2533**	0.1080	**0.1497**	**0.0668**	**0.0675**	**0.0319**	**0.3591**	**0.1811**
A^2-GCN	**0.2596***	**0.1166***	**0.1638***	**0.0766***	**0.0775***	**0.0350***	**0.3768***	**0.2019***
Improv.	2.49%	6.77%	9.42%	14.67%	14.81%	9.72%	4.93%	11.49%

The symbol * denotes that the improvement is significant with $p - value < 0.05$ based on a two-tailed paired t-test.

for testing.

Evaluation Metrics

For each user in the test set, we treated all the items that the user did not interact with as negative items. Two widely used evaluation metrics for top-n recommendation are adopted in our evaluation: Hit Ratio (HR) [30] and Normalized Discounted Cumulative Gain (NDCG) [52].

- **Hit Ratio (HR) [30]**: It is a recall-based metric, measuring whether the test item is in the top-n positions of the recommendation list (1 for yes and 0 otherwise).
- **Normalized Discounted Cumulative Gain (NDCG) [52]**: This metric emphasizes the quality of ranking, which assigns higher score to the top-ranked items by considering the position of correctly recommended.

For each metric, the performance is computed based on the top 20 results. Notice that the reported results are the average values across all the testing users.

Baselines

To demonstrate the effectiveness, we compared our proposed A^2-GCN with a set of strong competitors based on different approaches. Among the competitors, BPR-MF [5], NeuMF [6] and NGCF [7] only use the user-item interactions without considering

[5] https://github.com/Foolius/recsyslab/tree/master/bin/recommender

[6] https://github.com/hexiangnan/neural_collaborative_filtering

[7] https://github.com/xiangwang1223/neural_graph_collaborative_filtering

any side information; NFM [8], Wide&Deep [9], ACCM [10] and ENSFM [11] leverage the attribute information. We briefly introduced those baselines below.

- **BPR-MF[76]**: It is a classic matrix factorization based method for top-n recommendation, which only utilizes the user-item interactions and adopts the Bayesian personalized ranking loss for optimization.
- **NeuMF[56]**: It is a state-of-the-art neural collaborative filtering method, which employs multiple hidden layers above the element-wise and concatenation of user and item embeddings to capture their non-linear feature interactions. Especially, it uses two-layered plain architecture, where the dimension of each hidden layer keeps the same.
- **NGCF[136]**: This is a recently proposed GCN-based recommendation method. In particular, this method explicitly encodes the collaborative signal in the form of high-order connectivities by performing embedding propagation in the user-item bipartite graph. It has achieved the state-of-the-art performance among the methods which only use user-item interactions.
- **NFM[53]**: It is a deep neural factorization machine method, which uses Bi-Interaction Layer to integrate both attribute information and user-item interactions.
- **Wide&Deep[19]**: This method jointly trains a wide linear model with feature transformations and a deep neural network with embeddings. In the deep component, we used the same structure as reported the original paper.
- **ACCM[117]**: This is an attention-based model to unify CF-based and content-based recommendation for both warm and cold scenarios. This model can automatically choose proper attributes to represent the user and the item with the attention mechanism.
- **ENSFM[11]**: This method seamlessly connects the relationship between factorization machines and matrix factorization with non-sampling strategy.

To ensure fair comparisons, for all the methods using pair-wise learning, each positive instance is paired with a randomly sampled negative user-item instance in the training procedure. We put great efforts into tuning the hyperparameters of these methods and reported their best performance.

Parameter Tuning

We implemented our model with TensorFlow [12] and carefully tuned the key parameters. Specifically, we tuned the initial learning rate ℓ_0 among $\{0.1, 0.01, 0.001, 0.0001\}$. The coefficient of L_2 normalization is searched in $\{10^{-5}, 10^{-4}, \cdots, 10^1, 10^2\}$, and message dropout and node dropout are tuned in $[0, 0.8]$ with a step size of 0.1. We

[8] https://github.com/hexiangnan/neural_factorization_machine

[9] https://github.com/jrzaurin/pytorch-widedeep

[10] https://github.com/THUIR/ACCM/tree/master

[11] https://github.com/chenchongthu/ENSFM

[12] https://www.tensorflow.org.

optimized all models with the Adam optimizer, where the batch size is fixed at 1,024. In experiments, the dimension of latent vectors (*i.e.*, $\boldsymbol{e_u}$ and $\boldsymbol{e_v}$) of all methods is set to 64. We used the Xavier initializer [149] to initialize the model parameters. Besides, model parameters are saved in every 10 eopchs. We released the codes and involved parameter settings to facilitate others to repeat this work [13].

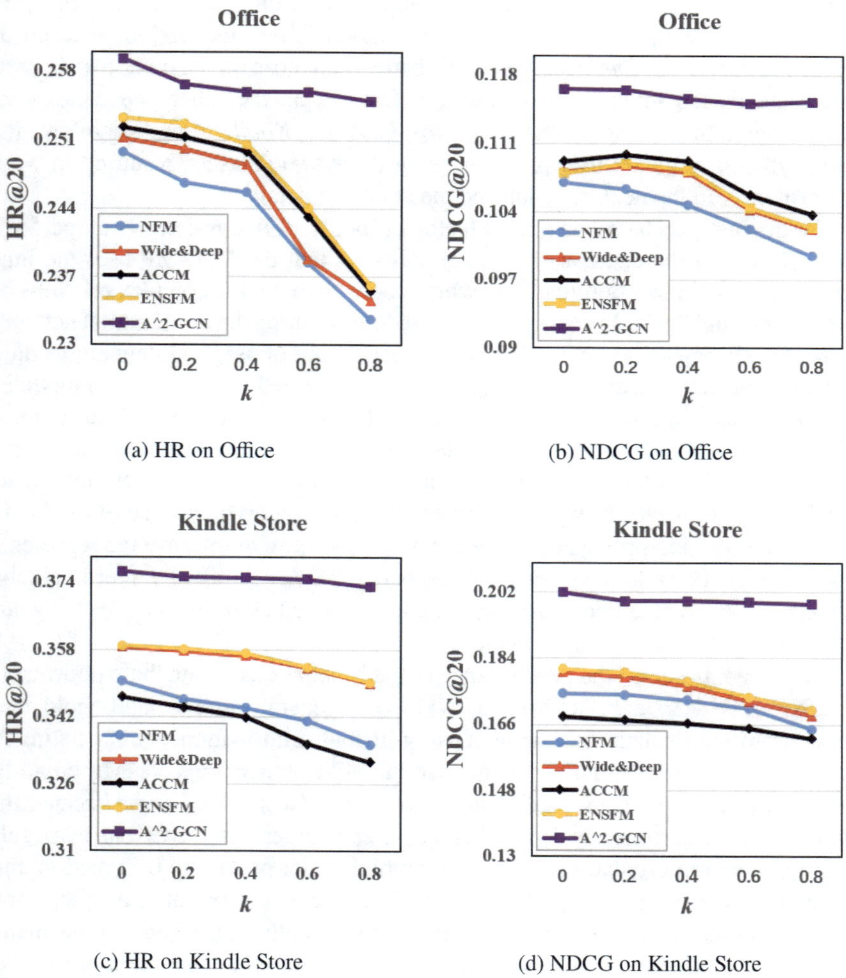

(a) HR on Office (b) NDCG on Office

(c) HR on Kindle Store (d) NDCG on Kindle Store

Fig. 5.4: Impact of attributes missing. Models are trained on training sets with all interactions and attribute labels are removed randomly in a ratio of {0, 0.2, 0.4, 0.6, 0.8}. k is the ratio of removed attributes in each dataset.

[13] https://github.com/liufancs/a2_gcn

5.5.2 Performance comparison (RQ1)

Table 5.4 reports the results of all involved methods. In the table, both the best and the second best performance is highlighted in bold. From the results, we obtained some interesting observations. In addition, we conducted pairwise significant test between our method and the baseline with the best performance.

Overall, A^2-GCN surpasses all the competitors consistently and significantly across all the cases, demonstrating its effectiveness on leveraging attributes for recommendation. Another interesting observation is that the performance on the *Office Products* and *Kindle Store* is much better than those on the other two datasets. The main reason might be that the item features on the other two datasets are more diverse than those on the *Office Products* and *Kindle Store*. Therefore, it is more difficult to model user preferences on those two datasets, resulting in worse performance. In the next, we analyzed the results in detail.

For the methods based on matrix factorization, NeuMF consistently outperforms BPR-MF on all the datasets. The main reason is that the BPR-MF uses the inner products as the interaction function, which cannot capture the complex relations between users and items. In contrast, NeuMF adopts multiple layers of neural networks above the element-wise interaction and concatenation of user and item embeddings to capture the non-linear interactions. This demonstrates the importance of modeling the complicated interactions between users and items. However, NGCF outperforms BPR-MF and NeuMF across all the datasets. This is attributed to the utilization of GCN techniques to learn user and item embeddings on the user-item interaction graph. In addition, NGCF exploits high-order connectivities between users and items through embedding propagation over the graph structure to improve the representation learning. For relatively denser datasets (*e.g.*, *Office Products*), more neighbor nodes (*e.g.*, interacted items) are available for each node's embedding learning, and thus better performance can be achieved.

Next, we introduced the performance of the baselines using attribute information (*i.e.*, NFM, Wide&Deep, ACCM and ENSFM). In general, these methods yield much better performance than those without using attribute information, demonstrating the great value of attributes for recommendation. NFM outperforms NeuMF on all the datasets but underperforms NGCF on the *Office Products*, which might because that *Office Products* is relatively denser. With the rich interactions, NGCF can learn solid user and item representations by leveraging high-order proximity information, thus achieving better performance. Wide&Deep yields better performance over the above methods on all the datasets. ACCM obtains consistently much better performance on *Office Products* and *Clothing*, owing to the adoption of the attention mechanism to control the ratio of information from different sources. This also indicates the importance of differentiating the influence of different information on user/item representation learning. However, ACCM underperforms NFM and Wide&Deep on *Toys Games* and *Kindle Store*. This is because the way of ACCM leveraging attributes makes it only deal with the limited number of attributes. In this chapter, we followed the setting of the ACCM in [117] to consider only the top 100 most frequent attribute labels, which causes great information loss. In contrast, NFM and Wide&Deep can tackle more attribute labels, and thus yield better performance on the two datasets. ENSFM surpasses the state-of-the-art methods using attribute information: NFM, Wide&Deep and ACCM. The reason is that non-sampling strategy can leverage all the data for computing gradient to learn the embedding.

A^2-GCN outperforms all the baselines consistently over all the datasets. In particular, compared to the strongest baseline in terms of NDCG@20, A^2-GCN can achieve a relative improvement by 6.77%, 14.67%, 9.72% and 11.49% on *Office Products*, *Toys Games*, *Clothing* and *Kindle Store*, respectively. The great improvement over NGCF demonstrates the effectiveness of leveraging attribute information in representation learning. Besides, our proposed method yields substantial improvement over the baselines (NFM, Wide&Deep, ACCM and ENSFM), which also exploit attribute information. This should be credited to the following reasons. Firstly, the GCN technique is used in our method to exploit the attribute information. Due to its powerful capability on representation learning, better performance is expected. Secondly, our model can naturally avoid the limitations of those baselines on dealing with the attribute missing problem, which also contributes to the performance improvement (see Section 5.5.3). Last but not least, our model benefits from the attention mechanism, which differentiates influence of attributes on user preference in the learning process (See Section 5.5.5). Note that A^2-GCN achieves the best performance on the first convolution layer on *Office Products* and *Clothing*. The reason is that when stacking more layers, the GCN-based models will suffer from the over-smoothing problem which leads the nodes embedding to become distinguishable. Besides, there could be noisy nodes (in higher-order neighbors) involved into the embedding propagation which also result in performance degradation. In particular, *Office Products* is a denser dataset, on which A^2-GCN more easily suffer from the above problems. And as discussed before, it is difficult to model users preferences on *Clothing* for its diverse item feature, which also results in worse performance when aggregating features of high-order neighbors.

Based on the above discussions, we yielded the following findings. The good performance of NeuMF reveals the importance of modeling non-linear features between users and items. The performance improvement achieved by NGCF on all datasets demonstrates the advantages of GCN-based models and the value of high-order information. Besides, NFM, Wide&Deep, ACCM and ENSFM indicate the effectiveness of leveraging attribute information to deal with the sparsity problem. Our A^2-GCN achieves the best performance because of its effectiveness on exploiting the attribute information via an attentive GCN method.

5.5.3 Effects of the Attribute Missing Problem (RQ2)

In this section, we studied the influence of the attribute missing problem on attribute-aware recommendation methods. To simulate the attribute missing problem, we randomly removed attribute labels from the datasets by a ratio in {0, 0.2, 0.4, 0.6, 0.8}. A ratio of 0.2 indicates that 20% of the attribute labels of items are randomly removed from the datasets. Due to the space limitation, we only presented the performance on the dataset *Kindle Store* and *Office Products*. The results on the other two datasets show similar trends.

We compared our model with the four baselines using attribute information. Fig. 5.4 shows the results in terms of HR@20 and NDCG@20 on two datasets. As we can see, A^2-GCN outperforms all the baselines by a large margin, especially when more attribute labels are missing, which indicates the effectiveness of our proposed method on modeling user preferences with attribute information. The performance of three baselines declines rapidly when more attribute labels are missing, especially for the *Office Products*. In particular, the performance of the three baselines deteriorates substantially when $k > 0.4$ on *Office Products*. According to statistics (see Table 5.3), the average number of attribute labels per item in *Kindle Store* and *Office Products* is 6.83 and 3.39, respectively. This may explain the dramatical performance drop on *Office Products*. On the one hand, the removal of attribute labels reduces information source for item and user representation learning, which equally affects all models. On the other hand, the strategies of those baselines for dealing with missing attributes introduce misleading information into the recommendation model, resulting in inferior performance.

The performance of our model is also negatively affected by the reduction of attribute information, but it is much more stable and the declination is much smaller, especially on the *Kindle Store* because of the relatively rich attribute information. This demonstrates that our model can exploit the attribute information in a much more efficient way via the information propagation through the graph structure. Besides, it also verifies the superiority of our model on dealing with the missing attribute problem over the baselines.

5.5.4 Effects of Data Sparsity (RQ3)

Table 5.5: Performance of our A^2-GCN model and its variants over four datasets.

Datasets	Metrics	GCN$_b$	A-GCN$_{am}$	A-GCN$_{att}$	A^2-GCN$_v$	A^2-GCN
Office Products	HR@20	0.2498	0.2575	0.2546	0.2593	**0.2596**
	NDCG@20	0.1071	0.1151	0.1143	0.1157	**0.1166**
Toys Games	HR@20	0.1347	0.1505	0.1561	0.1591	**0.1638**
	NDCG@20	0.0616	0.0682	0.0711	0.0731	**0.0766**
Clothing	HR@20	0.0603	0.0707	0.0717	0.0762	**0.0775**
	NDCG@20	0.0274	0.0323	0.0332	0.0343	**0.0350**
Kindle Store	HR@20	0.3409	0.3522	0.3631	0.3717	**0.3768**
	NDCG@20	0.1675	0.1844	0.1953	0.1981	**0.2019**

As important side information, attributes can help alleviate the data sparsity problem. To demonstrate the capability of A^2-GCN for users with limited interactions, we conducted experiments to study the performance of our method and other competitors over user groups with different sparsity levels. In particular, for each dataset, we clustered the users into four groups based on their interaction levels in the training data. Taking dataset *Clothing* as example, users are divided into four groups based on the number of interactions in the training dataset: less than 5, 10, 15, and more than 15. Fig. 5.5 shows the performance in terms of HR@20 on different user groups on dataset *Clothing* and *Kindle Store*. These figures also show the number of users in different groups. It can be seen that most users have less than 10 interactions in *Clothing* and *Kindle Store*, and some even with less than 5 interactions in *Clothing*, which further validates the common sparsity problem in real-world datasets. From the results, we obtained the following observations. ENSFM, ACCM, NFM and Wide&Deep consistently outperform all other baselines. This demonstrates that attribute information is indeed very useful in alleviating sparsity problem. In addition, NGCF surpasses NeuMF and BPR-MF, validating the effectiveness of exploiting collaborative signals from high-order connectivities. A^2-GCN yields significantly better performance than all the other baselines over all user groups, which verifies the robust capability of our model on exploiting attribute information.

Based on the further analysis of performance improvement on different groups over the two datasets, we obtained that the improvement of our model over that of the best baseline becomes larger with increasing interactions. Notice that our model benefits from both the attribute information and the GCN technique, which leverages the graph structure to learn user and item representations. With more connections (interactions) in the graph, our model can leverage the graph structure to better exploit the attribute information. It is also interesting to find that our model obtains better performance in the third group over the best baseline than that of other groups in *Kindle Store*. Specifically, relative improvements of A^2-GCN over the second best method ENSFM on the four groups are 3.72%, 7.99%, 10.43% and 6.14%, respectively. The reason might be that, when the interactions are relatively rich, the benefits from attribute and neighbors become relatively limited since interactions provide quit sufficient information for representation learning.

5.5.5 Ablation Study (RQ4)

In this section, we examined the contribution of different components to the performance of our model. Our analyses are based on the performance comparisons to following variants.

- **GCN**$_b$: In this model, attribute information and attention mechanism are removed from our method. It is a baseline model which merely applies the GCN technique to the user-item bipartite graph.
- **A-GCN**$_{am}$: This variant is designed to investigate the effectiveness of the attention mechanism without attribute information.

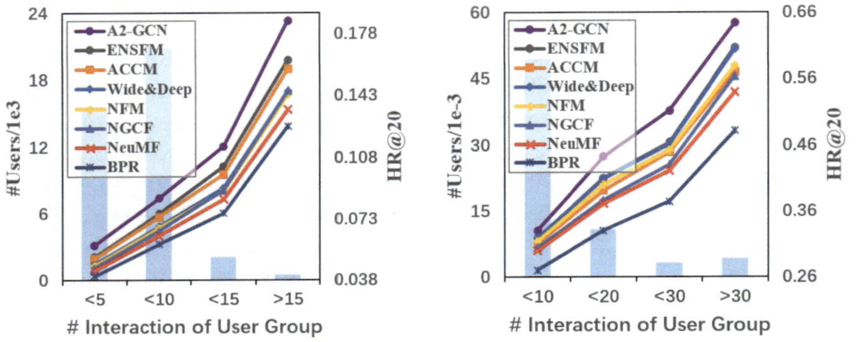

(a) Clothing (b) Kindle Store

Fig. 5.5: Performance comparison over the sparsity distribution of user groups on different datasets. Wherein, the background histograms indicate the number of users involved in each group, and the lines demonstrate the performance w.r.t. HR@20.

- **A-GCN$_{att}$**: This model removes the attention mechanism from our model to investigate the effects of attributes without attention mechanism.
- **A^2-GCN$_v$**: This model replaces the attribute-aware attention mechanism described in Eq. 5.12 via the common mechanism described in Eq. 5.10, where the attribute information has not been integrated into the attention mechanism. This variant is designed to study the effectiveness of our proposed attribute-aware attention mechanism.

The results of those variants and our method are reported in Table 5.5. A-GCN$_{am}$ outperforms GCN$_b$ over all the datasets, which indicates the importance of differentiating the information distilled from different neighbor nodes in GCN. From the perspective of recommendation, this validates the varying preferences of a user towards different items. The performance of A-GCN$_{att}$ indicates the effectiveness of leveraging attribute information for recommendation. In other words, the attributes can provide valuable information in learning more robust user and item representations. Although A^2-GCN$_v$ adopts a simplified attention mechanism, it consistently and substantially outperforms the A-GCN$_{att}$. This also demonstrates the effectiveness of the attention mechanism. More importantly, our A^2-GCN further improves the performance based on the proposed attribute-aware attention mechanism validating the assumption that items' attribute information affects a user's preference for this item. In addition, it also indicates our designed attention model can effectively capture this effect.

5.6 Conclusion

In this chapter, we present a novel model called Attribute-aware Attentive Graph Convolution Network (A^2-GCN), which can effectively exploit attribute information to learn user and item representations. More importantly, our model can naturally avoid the limitations of previous attribute-aware recommendation methods on dealing with the attribute missing problem. The GCN technique is adopted in A^2-GCN to model the complicated interactions among <users, items, attributes>. In particular, a novel attribute-aware attention mechanism is proposed to capture the effects of item attribute information on user preference. The experimental results on four real-world datasets demonstrate that our model can achieve substantially higher recommendation accuracy over several state-of-the-art methods. Additional experiments also validate the effectiveness of A^2-GCN on tackling the attribute missing problem and alleviating the sparsity problem.

Chapter 6
Light Graph Transformer Model

6.1 Introduction

A notable challenge in multimedia recommendation systems is feature misalignment, where extracted multi-modal features fail to align with user preferences or the specific objectives of recommendation tasks. To address this issue, GCNs can be leveraged to refine multi-modal features by incorporating user-item interaction data into the feature learning process. By propagating and aggregating information across graph structures, GCNs can filter out irrelevant signals and emphasize task-relevant features, aligning extracted content features more closely with recommendation objectives. In this chapter, we focus on tackling the feature misalignment challenge in multimedia recommendation. We explore how GCNs can refine extracted features, ensuring they better capture user preferences and contribute to more accurate recommendations.

Multimedia recommender system aims to discover the user preference for content information, in order to enrich the collaborative signal based on the collaborative filtering (CF) algorithm [112, 136, 87, 134, 92, 24, 89, 90]. Hence, it is becoming increasingly prevalent in online applications, ranging from search engines, E-commerce, to social media platforms.

To predict the user preference from the items' content information, the prior studies [51, 14, 144, 33] mostly follow a similar pipeline: the multi-modal feature extraction, user preference modeling, and user-item similarity measurement. Although these models introduce some advanced techniques to supercharge the recommendation, such as deep neural networks [14] and graph convolutional networks (GCNs) [144], they seldom explore the extracted features, which is the foundation of the user preference modeling and subsequent interaction prediction.

Probing the existing models, we argue that the extracted features contain much information irrelevant to the recommendation task, thereby failing to sufficiently model the user preference from the item content [84]. For visual inputs, early approaches primarily employed CNN-based pre-trained models as visual encoders. These visual encoders extract and compress visual features from raw pixel data

F. Liu, L. Nie, *Advancing Recommender Systems with Graph Convolutional Networks*,
https://doi.org/10.1007/978-3-031-85093-6_6

through convolution and pooling operations. For example, MMGCN [144] adopts ResNet [50], whereas VECF [18] uses VGG [121] as the visual encoder. Recently, Transformer-based visual pre-trained models have demonstrated superior performance [74, 160, 49, 93], leading to their adoption as visual encoders in multimodal recommender systems. For instance, ODMT [68] employs ViT [74] as its visual encoder. For textual modality, such as user reviews and item description, multimodal recommendation methods typically utilize feature extraction techniques such as Word2Vec [105], RNN [58] and BERT [31]. Taking the visual features extracted by the pre-trained ResNet [50] as an example, the discriminative signal tends to be emphasized to distinguish the images from different categories. Nevertheless, such a signal may pose an insignificant role for the user preference modeling. For instance, the features of shoelaces could help the classifier to differ the sneaker from dress shoes, whilst their importance will fall behind in the predictions of user-item interactions.

To overcome this problem, it is of great importance to distill the informative cues from off-the-shelf features for modeling user preference. The existing studies leverage the attention mechanism [14, 25] or deep learning method [51] to distill the informative signal by explicitly or implicitly reconstructing user-item historical interactions. Nevertheless, they ignore the correlation between the items observed by the same user in the features distillation, which reveals the users' interest in the content information. This is consistent with the assumption that similar items are prone to be recommended to the same users. That is, some features co-occurred in the items that reflect the user's tastes in their content information. Therefore, establishing the correlation between the items observed by the same user is able to discover the informative features that affect the user's decision-making.

Towards this end, we resort to Transformer techniques [125] to model the correlation between the items observed by the same user. Although recent years have witnessed many successful Transformer variants in the sequential and session recommendation [159, 176, 108], it is non-trivial to directly inherit the Transformer model in our work, due to two technical challenges:

- The extracted features of items, as the inputs of Transformer models, are dependent on pre-trained extractors, and thus some irrelevant information may harm the correlation modeling of item pairs. Therefore, *how to effectively measure the affinities between the items upon the off-the-shelf features is the first challenge to conduct the Transformer-based model.*
- Unlike the computer vision and natural language processing tasks, the input tokens in recommender system will be the exponential combination of the numerous items. Hence, *facing the overload and diversity inputs, how to efficiently perform the Transformer-based model is the other challenge in deploying the model.*

To address the challenges, we propose a Light Graph Transformer model (LightGT) to establish the item-item correlation, so as to learn the superior representation of user preference from the interacted items' features. Thereinto, we develop a modal-specific embedding and a layer-wise position encoder for the effective similarity measurement, and present a light self-attention block to improve the efficiency of self-attention scoring. Specifically, we gather the user with her/his interacted items together and embed them in each modality. These embeddings help to distill the users' interested features under the supervision of historical user-item interactions. Moreover, we gain the layer-wise position encodings based on the structure information of user-item graph to enrich the tokens' representations. It can help the attention scoring by introducing the CF signal into the self-attention measurement. In the light self-attention block, we empirically remove the multi-head attention, feed-forward network (FFN), and residual connection from the vanilla model for the efficiency. Owing to these designs, we can effectively and efficiently distill the informative features and learn the user preference on content information. Ultimately, we predict the user-item interactions by combining their CF- and content-based similarities.

To justify our proposed model, we conduct extensive experiments on three datasets, including Movielens, Tiktok, and Kwai datasets. The results demonstrate that the LightGT outperforms the state-of-the-art baselines by a significant margin. And interestingly, under the same experiment setting, the light self-attention block not only boosts the model's efficiency but leads to accuracy improvements.

In a nutshell, our main contributions are three-fold:

- After probing the prior studies, we propose a new Light Graph Transformer model (LightGT) for the multimedia recommendation. To the best of our knowledge, this is the first Transformer-based model for the multimedia recommendation.
- Within our proposed LightGT model, we develop the modal-specific embedding, layer-wise position encoder, and light self-attention block, in order to efficiently and effectively optimize the user and item representations for the predictions of user-item interactions.
- Conducting extensive experiments on three datasets, we demonstrate the effectiveness and efficiency of our proposed model. The insight we observed is that the devised light Transformer structure not only decreases the time complexity but improves the performance of recommendation.

6.2 Related Work

6.2.1 Multimedia Recommender System

In terms of multimedia recommender systems, pioneer researchers are prone to incorporating CF-based methods with content-based recommendation models [168, 59, 142, 86, 83, 12, 132, 175, 133, 143, 32, 91, 81]. Through restoring user-item interactions in the history, they target modeling the user preference on item contents to generate a high-quality recommender system. Various types of side information have been utilized in recommender systems, such as attributes [19, 15], reviews [103, 124, 22, 26] and images [104, 131]. For instance, He *et al.*, proposed Visual Bayesian Personalized Ranking (VBPR) approach [51], which extends BPR method and models the user preference on the visual information. After that, Chen *et al.,* developed an attention-based recommendation model, termed Attentive Collaborative Filtering (ACF) [14], to learn the user interests on both item- and component-level in the multimedia recommendation. Similarly, Liu *et al.* [95] presented a User-Video Co-Attention Network (UVCAN) for micro-video recommendation tasks, in order to capture multi-modal information from both the user and micro-video side. Considering the users' different tastes in the different modalities, Wei *et al.* [144] introduced GCNs into the multimedia recommendation model and proposed a Multi-modal Graph Convolution Network. Wherein, it is able to capture the model-specific user preference and optimize the item representations simultaneously. Based on the GCN structure, GRCN [143] utilizes the rich multimodal content of items to refine the structure of the interaction graph in order to mitigate the effect of false-positive edges on recommendation.

Although these methods achieve state-of-the-art results, they overestimate the importance of feature extraction in the multimedia recommendation. In this work, we consider the correlation between the items within the user's historical interaction and distill the item features captured from the pre-trained extractors, so as to effectively model the user preference for content information.

6.2.2 Transformer-based Recommendation Model

Recent years have witnessed the prevalence and effectiveness of the Transformer model in various domains ranging from natural language processing [73, 41, 40] to computer vision [74, 66, 67]. Taking advantage of the contextualized information modeling, the Transformer model recently draw much research attention from the recommendation domain to explore the relation among items. For instance, the prior studies [122, 17, 146, 150, 151, 39] construct the user-item interactions in history as the sequential structure and feed them into the Transformer-based recommendation methods to model user behavior sequences. Different from these sequential recommendation models, Liu *et al.,* [98] proposed a novel pre-trained multimodal graph transformer model to learn item representations by considering both item side information and their relationships. More recently, Li *et al.,* [80] developed a novel transformer-based model, named PEAR, for re-ranking. It not only captures interactions at feature- and item- levels, but models item contexts from the historically clicked item list.

However, these methods mostly inherit the designs of the vanilla Transformer model and leave their effectiveness and efficiency untouched. In our proposed model, we explore these operations and devise a light Transformer architecture for the multimedia recommendation.

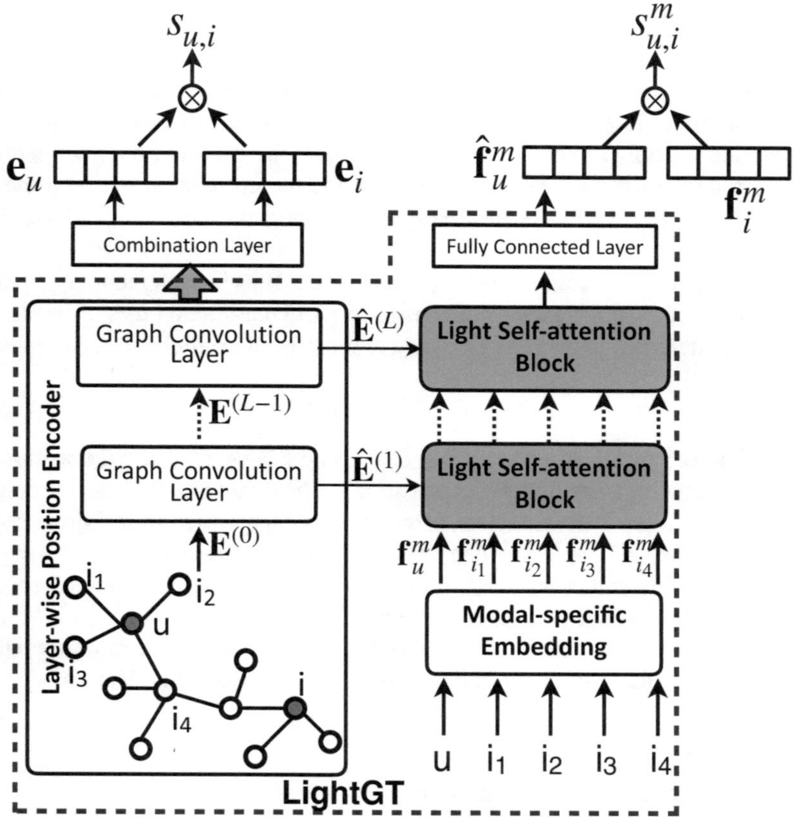

Fig. 6.1: An illustration of our proposed model.

6.3 LightGT Model

In the LightGT model, we extend the vanilla Transformer model by equipping with: (1) the modal-specific embedding that encodes the inputs of LightGT model in different modalities, (2) layer-wise position encoder which enriches the tokens with their roles in the historical interactions, and (3) light self-attention block that can be stacked to model the correlation between the tokens for user preference modeling. By performing the LightGT model in different modalities, we can learn the modal-specific user preference on the content information for user-item interaction prediction.

6.3.1 Modal-specific Embedding

In the Transformer-based model, it is essential to construct the input tokens for self-attention scoring. Here, our LightGT model targets at discovering the user preference from her/his historical interactions. We hence take the user and her/his interacted items as tokens of the model. It not only builds the correlation between the co-interacted items but distills the informative signal under the supervision of user-item interactions.

Inspired by the prior study [144], we embed the tokens in each modality, in order to learn the modal-specific user preference. In particular, taking user u in Fig. 6.1 and her/his interacted items (*i.e.*, i_1, i_2, i_3, and i_4) as an example, we can represent them as:

$$\mathbf{F}_u^m = [\mathbf{f}_u^m, \mathbf{f}_{i_1}^m, \mathbf{f}_{i_2}^m, \mathbf{f}_{i_3}^m, \mathbf{f}_{i_4}^m], \tag{6.1}$$

where $\mathbf{f}_{i_1}^m \in \mathbb{R}^{d\times 1}$, $\mathbf{f}_{i_2}^m \in \mathbb{R}^{d\times 1}$, $\mathbf{f}_{i_3}^m \in \mathbb{R}^{d\times 1}$, and $\mathbf{f}_{i_4}^m \in \mathbb{R}^{d\times 1}$ are the features of item i_1, i_2, i_3, and i_4 in modality m, respectively. They are captured from the content information by a pre-trained feature extractor and compressed into d-dimensional vectors. Moreover, $\mathbf{f}_u^m \in \mathbb{R}^{d\times 1}$ denotes the trainable representation of user u in modality m. After gathering these vectors together, we yield the modal-specific tokens' representations $\mathbf{F}_u^m$ for user preference modeling.

For the sake of notational brevity, in what follows, we conduct the same operations on different modalities and omit indicator m without specific clarification.

6.3.2 Layer-wise Position Encoder

Typically, position information plays a vital role in the Transformer model, which provides the prior knowledge to enhance the attention measurement. Unlike the sequential inputs, *e.g.*, words in sentences and frames in videos, our obtained tokens are not aware of the position in the sequence. As such, we resort to use the structure information of the user-item graph to encode the positions of tokens, which reflects their relations in the historical user-item interactions.

Towards this end, we define the id embeddings of users and items, and then reorganize the user-item interactions in the history as a bipartite graph:

$$\mathcal{G} = \{\mathbf{E}, \mathbf{A}\}. \tag{6.2}$$

Whereinto, $\mathbf{E} \in \mathbb{R}^{(N+M)\times d}$ is the trainable embedding matrix for the user and item nodes. And, $\mathbf{A} \in [0, 1]^{(N+M)\times(N+M)}$ is the adjacency matrix, functioning as the edges of the bipartite graph. More specifically, an edge $\mathbf{A}_{ui} = 1$ indicates an observed interaction between user u and item i; otherwise $\mathbf{A}_{ui} = 0$.

According to the construction of the interaction graph, the structure information can reflect the affinities between the nodes in the recommendation scenario. As shown in Fig. 6.1, given a pair of item nodes located in the two-hop neighbors (*e.g.*, i_2 and i_3), we can find they are interacted by at least one user node. Compared with the node pairs being distant in the graph (*e.g.,* i_2 and i), the nodes with more common neighbors tend to have more similar representations. Therefore, we can leverage the representations of structure information learned from the graph to encode the position of corresponding nodes (*i.e.,* item and user).

To encode the positions, we employ the GCN model on the bipartite graph for the structure information modeling. In particular, by iteratively performing the stacked graph convolutional layers, we can pass the message from multi-hop neighbors into centric nodes and thus represent their structure information with the neighbors' embeddings. Taking the l-th layer as an example, we model the nodes' structure information by aggregating their neighbors' embeddings, formally,

$$\begin{aligned}\mathbf{e}_u^{(l)} &= \sum_{i\in\mathcal{N}_u}\frac{1}{\sqrt{|\mathcal{N}_i|}\sqrt{|\mathcal{N}_u|}}\mathbf{e}_i^{(l-1)},\\ \mathbf{e}_i^{(l)} &= \sum_{u\in\mathcal{N}_i}\frac{1}{\sqrt{|\mathcal{N}_i|}\sqrt{|\mathcal{N}_u|}}\mathbf{e}_u^{(l-1)},\end{aligned} \tag{6.3}$$

where $\mathcal{N}_u$ and $\mathcal{N}_i$ denote the sets of neighbor nodes connecting to user u and item i, respectively. $|\mathcal{N}_u|$ and $|\mathcal{N}_i|$ are the number of nodes in two sets. The symmetric normalization term $\frac{1}{\sqrt{|\mathcal{N}_i|}\sqrt{|\mathcal{N}_u|}}$ is able to avoid the scale of embeddings when stacking the graph convolutional layers. $\mathbf{e}_u^{(l)}$ and $\mathbf{e}_i^{(l)}$ are the structure information learned from the l-hop neighbors of user u and item i. Note that we set $\mathbf{e}_u^{(0)}$ as the embedding of user u, *i.e.,* $\mathbf{e}_u \in \mathbf{E}$; analogically, $\mathbf{e}_i^{(0)}$ equals to $\mathbf{e}_i \in \mathbf{E}$.

Thereafter, we implement the layer-wise position encoder that absorbs the structure information and generates the position encodings for the different self-attention blocks, formally,

$$\begin{cases}\hat{\mathbf{E}}^{(1)} = \frac{1}{2}(\mathbf{E}^{(0)} + \mathbf{E}^{(1)})),\\ \hat{\mathbf{E}}^{(2)} = \frac{1}{3}(\mathbf{E}^{(0)} + \mathbf{E}^{(1)} + \mathbf{E}^{(2)})),\\ \qquad\cdots,\\ \hat{\mathbf{E}}^{(L)} = \frac{1}{L+1}(\mathbf{E}^{(0)} + \mathbf{E}^{(1)} + \cdots + \mathbf{E}^{(L)}),\end{cases} \tag{6.4}$$

where $\mathbf{E}^{(l)}$ denotes the representations of structure information learned from l-th graph convolutional layer. $\hat{\mathbf{E}}^{(l)}$ represents the position encoding for the l-th self-attention block. The motivation of this position encoder is that we need more sufficient structure information to help the self-attention scoring when we stack the self-attention blocks.

Furthermore, we consider the low confidence of tokens' representations at the beginning of the training phase, due to their irrelevant signal learned from the off-the-shelf extractors. To alleviate their negative feedback to the structural information modeling, we further introduce a stop-gradient operation into the position encoder, formally,

$$\hat{\mathbf{E}}^{(l)} = stopgrad(\hat{\mathbf{E}}^{(l)}) \tag{6.5}$$

where $stopgrad()$ represents the stop-gradient operation to cut off the back propagation from the self-attention blocks to the user and item embeddings.

Overall, from the perspective of recommendation algorithm, the layer-wise position encoder introduces the CF-based similarities between nodes into the self-attention blocks at different levels. With such a position encoder, we can optimize the self-attention measurement by incorporating the external knowledge of user-item interactions with tokens' features, in order to discover the features relevant to the user preferences.

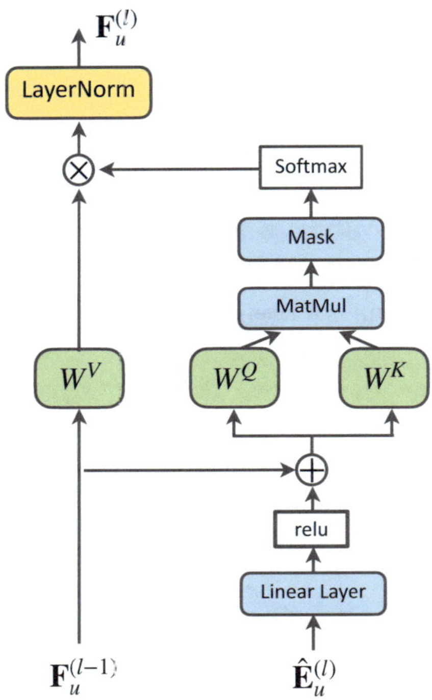

Fig. 6.2: An illustration of the l-th light self-attention block.

6.3.3 Light Self-attention Block

Recalling the prior studies, the promising performance of Transformer has been witnessed in various applications ranging from natural language processing to computer vision. However, we find that it is hard to optimize the standard Transformer efficiently and accurately, due to the data overload and sparsity interactions in the recommender system. After experimentally analyzing the components in the vanilla Transformer, we devise a light self-attention block by simplifying the heavy and burdensome components.

To be implemented, we first embed the tokens' position encodings $\hat{\mathbf{E}}_u^{(l)}$ into their representation. As Fig. 6.2 displays, in the l-th block, we formulate the position-enriched representations of tokens as:

$$\mathbf{H}_u^{(l)} = sigmoid(\mathbf{W}_m \hat{\mathbf{E}}_u^{(l)} + \mathbf{b}_m) \oplus \mathbf{F}_u^{(l-1)}, \tag{6.6}$$

where $\oplus$ represents the element-wise addition operation. $\mathbf{F}_u^{(l-1)}$ denotes the tokens' representations learned from the previous block. As for the first block, $\mathbf{F}_u^{(0)}$ is the off-the-shelf features of tokens, *i.e.*, $\mathbf{F}_u^m$. To fuse these two vectors, we project the position encoding to m modality with a linear layer equipped with the activation function $sigmoid()$, where $\mathbf{W}_m \in \mathbb{R}^{d\times d}$ and $\mathbf{b}_m \in \mathbb{R}^{d\times 1}$ are the trainable parameters in modality m. And then, we can achieve the position-enriched token representations at l-th layer, *i.e.*, $\mathbf{H}_u^{(l)}$.

With the enriched representations of tokens, we can obtain the query and key vectors for similarity measurement, formally,

$$\mathbf{Q} = \mathbf{W}^Q \mathbf{H}_u^{(l)}, \mathbf{K} = \mathbf{W}^K \mathbf{H}_u^{(l)}, \tag{6.7}$$

where $\mathbf{W}^K \in \mathbb{R}^{d\times d}$, and $\mathbf{W}^Q \in \mathbb{R}^{d\times d}$ are the trainable parameters to map the tokens' representations into the query and key spaces, respectively.

Next, we empirically remove the FFN from each self-attention block and residual connection between the blocks. More importantly, we find that the single-head attention module can achieve a comparative performance with the multi-head attention one. Therefore, we conduct the scaled self-attention operations:

$$\mathbf{F}_u^* = softmax(\frac{\mathbf{Q}\mathbf{K}^\top}{\sqrt{d}})\mathbf{W}^V \mathbf{F}_u^{(l-1)}, \tag{6.8}$$

where $\mathbf{F}_u^*$ is the tokens' representations enhanced by aggregating the context information. $\mathbf{W}^V \in \mathbb{R}^{d\times d}$ represents the parameter to gain the value vectors. It is worth noting that we instead use the tokens' representations without the position encodings (*i.e.*, $\mathbf{F}_u^{(l-1)}$). This is because that we focus on the user preference on the content information and avoid the effects from the id embeddings.

Thereafter, we employ the normalization operation and generate the tokens' representation at l-th self-attention block:

$$\mathbf{F}_u^{(l)} = LayerNorm(\mathbf{F}_u^*), \tag{6.9}$$

where $\mathbf{F}_u^{(l)}$ is the output matrix of tokens at l-th block and $LayerNorm()$ denotes the layer normalization function.

Taking advantages of the self-attention mechanism, we model the relation of user-item and item-item pairs. Thereinto, the user-item affinities can help to emphasize the items' features users prefer, and the item-item correlation highlight the co-occurrence information affecting the user's decision-making.

6.3.4 User preference Modeling

Stacking L self-attention blocks, we obtain the enhanced representations of input items and user tokens. For the users, some informative cues are captured at different blocks and aggregated with their representations. To model the user's preference on the item content, we formulate a fully connected network (FCN) to re-express user u:

$$\hat{\mathbf{f}}_u = leaky\_relu(\mathbf{W}^{'}_m \mathbf{f}_u^{(L)} + \mathbf{b}^{'}_m), \tag{6.10}$$

where $\mathbf{f}_u^{(L)} \in \mathbf{F}_u^{(L)}$ is u's representation after the L-th light self-attention block. $\mathbf{W}^{'}_m$, $\mathbf{b}^{'}_m$, and $leaky\_relu$ are the projection matrix, bias vector, and Leaky_relu function [100], respectively. Ultimately, we obtain the enhanced representation of user preference $\hat{\mathbf{f}}_u$.

6.3.5 Model Prediction

In order to predict the interactions between users and items, we follow the common solution of multimedia recommendation [51, 14, 144] and separately estimate their CF-based and content-based similarities.

In particular, we are able to achieve the CF-based embeddings of users and items based on the graph convolutional networks. By combining the structure information (*a.k.a.* collaborative signal) learned from L layers, we can inject the high-order interaction information into the embeddings [54]. We formulate the combination layer as:

$$\mathbf{e}_u = \sum_{l=0}^{L} \mathbf{e}_u^{(l)}, \mathbf{e}_i = \sum_{l=0}^{L} \mathbf{e}_i^{(l)}. \tag{6.11}$$

Then, we can measure the similarities between users and items upon their CF-based representations, as,

$$s_{u,i} = \mathbf{e}_u \mathbf{e}_i^{\top}, \tag{6.12}$$

where $s_{u,i}$ is our desired score estimating how likely user u prefers item i according to the historical behaviours.

As for the content-based score, we gain the item features from the off-the-shelf features and calculate the similarity between the user and target item on modality m, formally,

$$s_{u,i}^m = \hat{\mathbf{f}}_u^m (\mathbf{f}_i^m)^{\top}. \tag{6.13}$$

Whereinto, $s_{u,i}^m$ scores how user u prefers the content of item i.

With the obtained CF- and content-based scores, we can combine them to predict the user-item interaction, formally,

$$s^*_{u,i} = \lambda s_{u,i} + (1-\lambda) \sum_{m \in \mathcal{M}} s^m_{u,i}, \tag{6.14}$$

where λ is a hyper-parameter to balance the two types of information in the final interaction score $s^*_{u,i}$.

6.3.6 Optimization

To optimize the parameters of our proposed model, we apply Bayesian Personalized Ranking (BPR) [112], which is a well-known pairwise ranking optimization algorithm, and reconstruct the historical interactions between users and items.

Towards this end, we organize the triplet of one user, one item she/he observed, and one item unobserved by the user. It can be defined as,

$$\mathcal{R} = \{(u,i,i')|(u,i) \in \mathcal{O}, (u,i') \notin \mathcal{O}\}, \tag{6.15}$$

where $\mathcal{R}$ is a set of triples for training. According to the assumption that the user prefers the observed item rather than the unobserved one, the objective function of BPR can be formulated as,

$$\mathcal{L} = \sum_{(u,i,i') \in \mathcal{R}} -\ln \varphi(s^*_{u,i} - s^*_{u,i'}) + \eta \left\| \Theta \right\|_2^2, \tag{6.16}$$

where $\varphi(\cdot)$, η, and Θ denote the $sigmod$ function, regularization weight, and trainable parameters of our proposed model, respectively.

6.4 Experiment

To evaluate the effectiveness of our proposed model, we conduct extensive experiments on three public datasets and answer the following research questions:

- **RQ1** How does our proposed model perform compared with state-of-the-art multimodal recommendation models?
- **RQ2** How do the designs (*i.e.* layer-wise position encoder and light self-attention block) in LightGT affect the performance of our model?
- **RQ3** What is the effectiveness and efficiency of the components (*i.e.,* multi-head attention, FFN, and residual connection) in the self-attention block?
- **RQ4** Can LightGT learn the superior representation of user preference?

Before answering the above four questions, we describe the datasets, evaluation protocols, baselines, and parameter settings in the experiments.

6.4.1 Experiment Settings

Table 6.1: Statistics of the evaluation datasets. (d_v, d_a, and d_t denote the dimensions of visual, acoustic, and textual modality feature data, respectively.)

Dataset	#Users	#Items	#Inter	d_v	d_a	d_t
MovieLens	55,485	5,986	1,239,508	2048	128	100
TikTok	36,656	76,085	726,065	128	128	128
Kwai	7,010	86,483	298,492	2048	-	128

Dataset

To test the effectiveness of our proposed model, we follow the prior studies and conduct extensive experiments on three datasets, including Movielens, Tiktok, and Kwai, which are widely used in current research on multimedia recommendation. More detailed information on these datasets could be found in Table 6.1.

- **Movielens** Movielens dataset[1] are designed for the studies of recommender systems. In terms of multimedia recommendation, the authors of MMGCN collected the videos' trailers, titles, and textual descriptions. With the pre-trained models (*e.g.* ResNet [50], VGGish [57], and Sentence2Vector [1]), the visual, acoustic, and textual features are extracted from the frames, audio tracks, and descriptions, respectively.
- **Tiktok** The items in this dataset contain multimedia features, involving visual, acoustic, and textual modalities. According to the publisher of Tiktok[2], the features are extracted by some deep learning models pre-trained on non-recommendation datasets.
- **Kwai** Similar to the above two datasets, the items in the Kwai dataset consist of multi-modal features obtained by the feature extractors. The extractors are also designed for the recognition tasks on different modalities and trained with datasets irrelevant to recommendation tasks.

Evaluation Protocols

For each dataset, we randomly split the interaction records per user into the training, validation, and testing sets under the ratio $8:1:1$. The validation set and testing set

[1] https://movielens.org/.
[2] https://www.tiktok.com/.

are respectively used to tune the hyper-parameters and evaluate the performance in the experiments. Moreover, following the widely-used evaluation metrics [136, 54], we adopted recall@K (R@10) and Normalized Discounted Cumulative Gain (N@K) to evaluate the performance of methods. By default, we set $K = 10$ and reported the average values of the two metrics for all users in the test set. We make use of Nvidia Tesla V100 graphics card (32GB Memory) for all of the experiments.

Baselines

We compare our proposed model with the state-of-the-art baselines, including the CF-based recommendation models (*i.e.*, *GraphSAGE* [3], *NGCF* [4], *GAT* [5], and *LightGCN* [6]) and multimedia recommendation models (*i.e.*, *VBPR* [7], *MMGCN* [8], *GRCN* [9], and *LATTICE* [10])[11]. In particular,

- **GraphSAGE** [48]: This model conducts the graph convolutional operations on the user-item graph and injects the structure information of nodes into their representations, in order to optimize the predictions of user-item interactions.
- **NGCF** [136]: This model presents a recommendation framework to explicitly integrate the CF signal into the user and item embeddings. By exploiting the high-order connectivity from user-item interactions, the model encodes more effective CF signals into the representations.
- **GAT** [126]: The method aims to learn the weights for user-item interaction, *i.e.,* edges in the user-item graph, according to the similarity of node pairs. It adaptively aggregate the neighbors' messages of each node and enhance the node's representation for interaction predictions.
- **LightGCN** [54]: LightGCN is a popular GCN-based recommendation model. By removing the feature transformation and nonlinear activation, a light graph convolution layer is designed to effectively and efficiently model the high-order interaction information into the user and item embeddings.
- **VBPR** [51]: This is a benchmark model in the multimedia recommendation. It incorporates multi-modal features extracted by pre-trained models with matrix factorization framework to predict the interactions between users and items.
- **MMGCN** [144]: MMGCN is the first GCN-based multimedia recommendation model, which constructs the user-item graph in each modality and learns the modal-specific user preference with graph convolutional operations. It directly

[3] https://github.com/williamleif/GraphSAGE

[4] https://github.com/xiangwang1223/neural_graph collaborative_filtering

[5] https://github.com/PetarV-/GAT

[6] https://github.com/kuandeng/LightGCN

[7] https://github.com/arogers1/VBPR

[8] https://github.com/weiyinwei/MMGCN

[9] https://github.com/weiyinwei/GRCN

[10] https://github.com/CRIPAC-DIG/LATTICE

[11] Here, we omit ACF model, since the raw data, like keyframes and soundtrack, is not provided in the Kwai and Tiktok datasets.

takes the extracted features as items' representation and accordingly learns the user preference from the historical items.

- **GRCN** [143]: This model focuses on refining the structure of the user-item graph to alleviate the bias caused by false positive interaction records. Hence, the model leverages the extracted features to measure the similarities between users and items, and then effectively learns user preference based on the off-the-shelf features.
- **LATTICE** [167]: This method is designed to model the correlation between the items by computing the cosine distances of their features, in order to enrich the CF-based signal with the semantic similarities for the recommendation.

Parameter Settings

We use the Pytorch[12] and torch-geometric packages[13] to implement our proposed model. In particular, we initialize the parameters of model with Xavier algorithm [149] and optimize them with Adam optimizer [70]. For the learning rate and regularization weight, we conduct the grid search in $\{0.0001, 0.001, 0.01, 0.1, 1\}$ and $\{0.00001, 0.0001, 0.001, 0.01, 0.1\}$, respectively. When recall@10 on the validation data does not increase for 20 successive epochs, we stopped the training and reported the results on testing dataset. For the baselines, we followed the designs in their articles to achieve the best performance. For fairness, we did the same options and fixed the dimension of the ID embedding vector to 64.

6.4.2 Performance Comparison (RQ1)

To demonstrate the effectiveness of our proposed model, we conduct the experiments on three datasets and list the results of LightGT and baselines in Table 6.2. According to the results, we have the following observations:

- On three datasets, our proposed LightGT substantially outperforms all the baselines by a large margin. In particular, LightGT improves the strongest baselines *w.r.t.* Recall by 5.16%, 22.77%, and 8.77%, on the three datasets respectively. Such improvements verify the effectiveness of our proposed model. We attribute the performance to the items' informative features extraction based on their correlation modeling, so as to optimize the representations of user preference on the content information. Moreover, p-value ¡ 0.05 shows that the improvements in our model are statistically significant.
- LightGT and the strongest baseline LATTICE perform better than the other baselines in most cases. It indicates that explicitly modeling the correlation between items can help the prediction of user-item interaction. Further, LATTICE clearly underperforms our LightGT in terms of two metrics. It makes sense since LATTICE learns the user preference based on the raw features learned by the

[12] https://pytorch.org/.

[13] https://pytorch-geometric.readthedocs.io/.

Table 6.2: Overall performance comparison between our model and the baselines on three datasets.

Methods	MovieLens		Tiktok		Kwai	
	R@10	N@10	R@10	N@10	R@10	N@10
GraphSAGE	0.2129	0.1388	0.0778	0.0476	0.0424	0.0344
NGCF	0.2340	0.1383	0.0906	0.0547	0.0427	0.0363
GAT	0.2342	0.1589	0.0945	0.0575	0.0441	0.0369
LightGCN	0.2381	0.1592	<u>0.0988</u>	0.0603	<u>0.0502</u>	<u>0.0411</u>
VBPR	0.1927	0.1207	0.0600	0.0397	0.0302	0.0221
MMGCN	0.2453	0.1523	0.0645	0.0579	0.0456	0.0374
GRCN	<u>0.2520</u>	<u>0.1683</u>	0.0952	0.0584	0.0473	0.0403
LATTICE	<u>0.2520</u>	0.1680	0.0984	<u>0.0611</u>	0.0483	0.0400
LightGT	**0.2650**	**0.1771**	**0.1213**	**0.0751**	**0.0546**	**0.0441**
% Improv.	5.16%	5.23%	22.77%	22.91%	8.77%	7.30%
p-value	1.34e-4	3.01e-4	6.49e-7	8.51e-6	8.71e-5	1.84e-4

pre-trained extractor, while our LightGT captures the informative cues from the raw ones for user preference modeling.

- Compared with content-based models, CF-based recommendation models achieve poor performance in most cases. The reason for their suboptimal results might be the lack of the content-based similarity of the user-item pair. This verifies the reasonability of the motivation and designs of our proposed model again.
- Cross three datasets, it is easy to find that the improvements of LightGT on Tiktok and Kwai datasets are larger than that on Movielens. We suggest that this might be caused by their different densities, *i.e.,* interactions per user. Hence, the performance on the sparser datasets can be improved larger by the superior content-based similarity.

6.4.3 Ablation Study (RQ2)

Effect of Layer-wise Position Encoder

To test the layer-wise position encoder, we compare LightGT with several variant models:

w/o PE: We remove the position encoder and conduct the light self-attention blocks merely on the content features.

w/o SG: In this variant, we remove the stop-gradient operation from the layer-wise position encoder, *i.e.,* Eq. 6.5.

w/o LW: Removing the layer-wise designs, we calculate the position encoding in each block with the average of representations of structure information learned by all LGC layers.

Table 6.3: Overall performance comparison between our model and the baselines on three datasets.

Methods	MovieLens		Tiktok		Kwai	
	R@10	N@10	R@10	N@10	R@10	N@10
w/o PE	0.2604	0.1753	0.1192	0.0729	0.0513	0.0430
w/o SG	0.2599	0.1738	0.1074	0.0644	0.0531	0.0435
w/o LW	0.2627	0.1767	0.1207	0.0749	0.0530	0.0435
LightGT	0.2650	0.1771	0.1213	0.0751	0.0546	0.0441

According to the results listed in Table 6.3, we find that our proposed LightGT consistently achieves superior performance compared with three variant models. In particular, after removing the position encoder from the model (*i.e.,* **w/o PE**), the performance drops on three datasets, which verifies the effectiveness of position encoding in the Transformer-based model. Moreover, conducting the layer-wise position encoder without the stop-grained operations(**w/o SG**) degrades the performance clearly. The reason might be that the suboptimal item features bias the CF embeddings by passing the noisy signal during the back-propagation, especially at the beginning of the training phase. As for the variant model **w/o LW**, its inferior performance indicates that the encodings may be inconsistent in different blocks and empirically verifies the layer-wise design in the position encoder.

Effect of Light Self-attention Block

For the light self-attention block, we test the layer (*i.e.,* block) numbers in the range of $\{1, 2, 3, 4\}$. As a comparison, we conduct the LightGCN under the same setting.

As illustrated in Fig. 6.3, Fig. 6.4 and Fig. 6.5, we observe that our LightGT significantly outperforms LightGCN on three different datasets when the number of layers is larger than two. Furthermore, by increasing the layer number, the improvements achieved by LightGT are more significant than that of LightGCN. From the observations, we conclude the findings that: 1) the deeper GCNs in our model can model and inject more comprehensive structure information into the tokens' representations, further enhancing the self-attention scoring; and 2) stacked light self-attention blocks are capable of distilling the effective features from the off-the-shelf ones. This again verifies the effectiveness and reasonability of the designs in our proposed model.

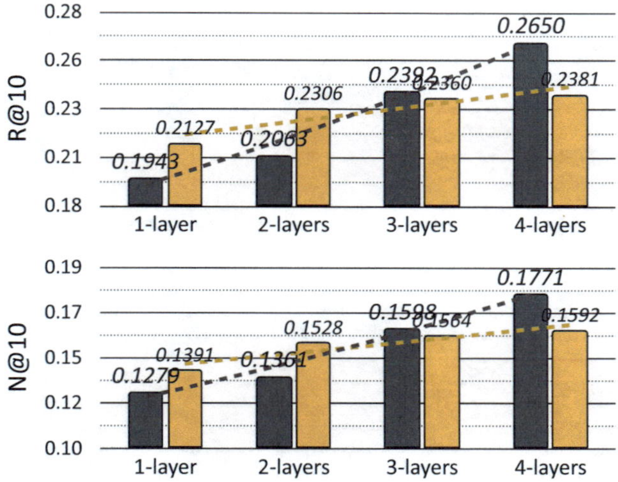

Fig. 6.3: Results of LightGT and LightGCN in terms of different layers on Movielens. (Green and yellow colors remark LightGT and LightGCN, respectively. Best viewed in color.)

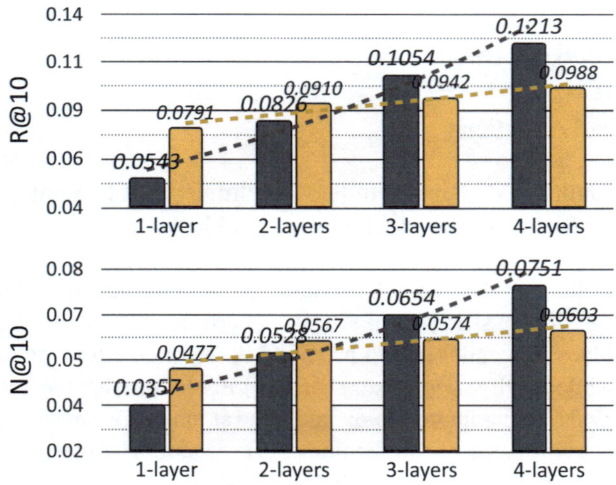

Fig. 6.4: Results of LightGT and LightGCN in terms of different layers on Tiktok. (Green and yellow colors remark LightGT and LightGCN, respectively. Best viewed in color.)

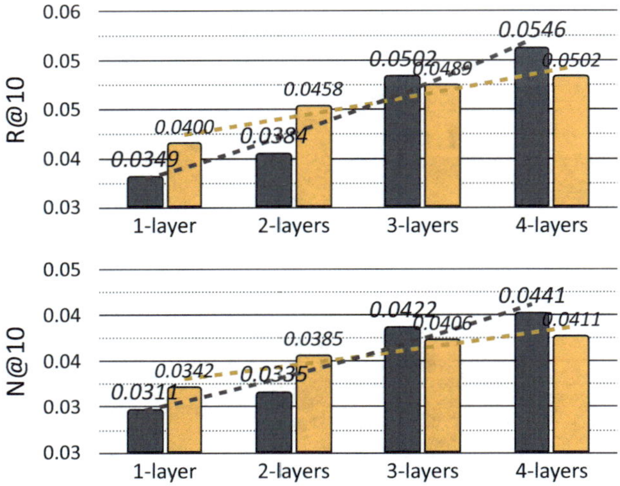

Fig. 6.5: Results of LightGT and LightGCN in terms of different layers on Kwai. (Green and yellow colors remark LightGT and LightGCN, respectively. Best viewed in color.)

6.4.4 In-depth Analysis (RQ3)

Effect of Multi-head Attention

To investigate the multi-head attention in the Transformer-based recommendation model, we compare 2-head and 4-head self-attention blocks with our LightGT on three datasets.[14]

We list their results *w.r.t.* R@10, N@10, and Times in Table 6.4 and have the observation that LightGT consistently gains the best performance on all the datasets. Specifically, the values of recall@10 and NDCG@10 decrease with increasing the number of heads, while the time cost raises sharply. It figures out that the multi-head self-attention is helpful in computer vision and natural language processing tasks, but will harm the performance of the recommender system. The reason might be that it is enough to model the tokens' correlation in a single aspect.

Effect of FFN

Considering the effectiveness of FFN in vanilla Transformer, we implement it in our light self-attention block to test its necessity in recommendation tasks. For this

[14] We only test 2- and 4-head self-attention variants due to memory limitations of the GPUs we used.

Table 6.4: Performance comparison between single- and multi-head self-attention blocks on three datasets. (Time (s) denotes the seconds per epoch. Head # represents the number of heads in the self-attention block.)

	Head #	R@10	N@10	Time (s)
Movielens	1 Head	0.2650	0.1771	167s
	2 Heads	0.2621 (-1.09%)	0.1753 (-1.02%)	229s
	4 Heads	0.2613 (-1.40%)	0.1741 (-1.70%)	334s
Tiktok	1 Head	0.1213	0.0751	80s
	2 Heads	0.1113 (-8.24%)	0.0695 (-7.46%)	106s
	4 Heads	0.1095 (-9.72%)	0.0683 (-2.24%)	164s
Kwai	1 Head	0.0546	0.0441	35s
	2 Heads	0.0515 (-5.68%)	0.0423 (-4.08%)	47s
	4 Heads	0.0523 (-4.21%)	0.0427 (-3.17%)	61s

Table 6.5: Performance comparison between LightGT and the baseline equipped with FFN (w/ FFN) on three datasets.

	Methods	R@10	N@10	Time (s)
Movielens	LightGT	0.2650	0.1771	167s
	w/ FFN	0.2474 (-6.64%)	0.1645 (-7.11%)	255s
Tiktok	LightGT	0.1213	0.0751	80s
	w/ FFN	0.1148 (-5.36%)	0.0709 (-5.59%)	154s
Kwai	LightGT	0.0546	0.0441	35s
	w/ FFN	0.0508 (-6.96%)	0.0424 (-3.85%)	79s

goal, we conduct the experiments on three datasets and record its results *w.r.t.* R@10 and N@10 as well as the time consumption in Table 6.5.

Observing the results, we surprisingly find that the performance largely drops when we implement FFNs in the self-attention block (*i.e.,* **w/ FFN**). We attribute it to the overfitting problem, since the FFNs introduce some parameters to be learned. More specifically, unlike the tasks in the computer vision and natural language processing domains, sufficiently modeling the semantic information might be unnecessary in the recommendation application. Jointly analyzing the time consumption of the two models, we easily find that the light self-attention block benefits both the effectiveness and efficiency of our Transformer-based model.

Effect of Residual Connection

To validate the residual connection, we conduct the experiments on three datasets by equipping the residual connection between two successive self-attention blocks, *i.e.,* **w/ RC**. Beyond the recall, and NDCG metrics, we report the time cost of each epoch.

From the results shown in Table 6.6, we observe that LightGT achieves comparative performance *w.r.t.* R@10 and N@10 with the baseline model. Even without the residual connection, unexpectedly, our LightGT outperforms the baseline on three

Table 6.6: Performance comparison between LightGT and the baseline equipped with residual connection (w/ RC) on three datasets.

	Methods	R@10	N@10	Time (s)
Movielens	LightGT	0.2650	0.1771	167s
	w/ RC	0.2606 (-1.66%)	0.1750 (-1.19%)	188s
Tiktok	LightGT	0.1213	0.0751	80s
	w/ RC	0.1135 (-6.43%)	0.0705 (-6.13%)	112s
Kwai	LightGT	0.0546	0.0441	35s
	w/ RC	0.0496 (-9.15%)	0.0405 (-8.16%)	67s

datasets by a margin. More importantly, compared with the baseline, we find that LightGT costs significantly less time per epoch, which comes from removing residual connections. Overall, the observation verifies that our design not only has a slight effect on the accuracy but improves the efficiency of model training and inference.

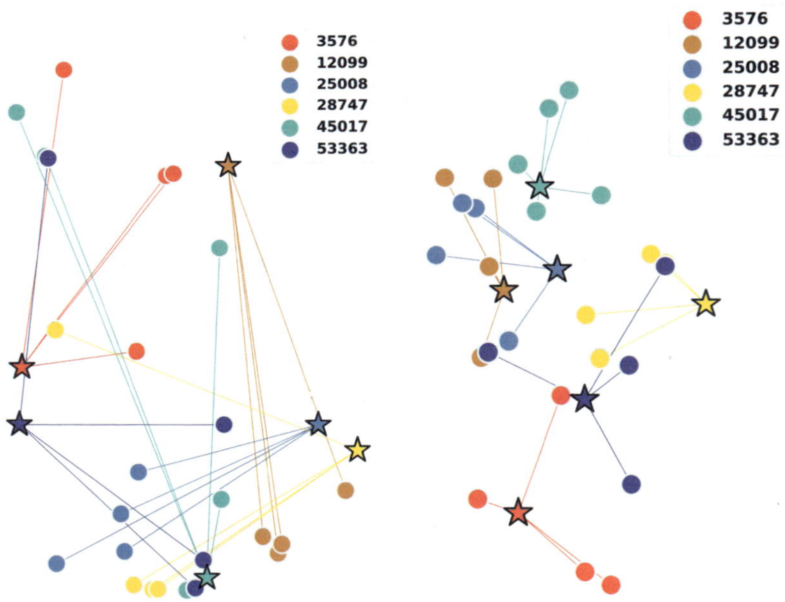

(a) MMGCN (b) LightGT

Fig. 6.6: Visualization of the t-SNE representations of user and items learned from MMGCN and LightGT. (Stars denote the users and the points with the same color are the relevant items. Best viewed in color.)

6.4.5 Visualization (RQ4)

In this section, we explore whether the representations derived from our proposed model are better to measure the user preference for the items' content information. Toward this end, we randomly selected six users from the Tiktok dataset. After collecting several items that they are never paired in the training phase, we conduct the t-SNE algorithm to plot their representations on a two-dimension space and visualize the item and user representations learned from MMGCN and LightGT models, as shown in Fig. 6.6.

Compared with the MMGCN model, we observe that the stars and points with the same color, *i.e.,* the user and her/his relevant items, are embedded nearer when performing the LightGT model. These discernible clusters derived from the LightGT model show that our model effectively model the user preference. We attribute it to that LightGT model is capable of capturing the informative signal from the items' features. It verifies the effectiveness of our proposed model again.

6.5 Conclusion

In this chapter, we propose a new Transformer-based model, named Light Graph Transformer (LightGT), for multimedia recommendation. Within the proposed model, we design a modal-specific embedding and a layer-wise position encoder for effective features distillation, and devise a light self-attention block for efficient self-attention scoring. By modeling the correlation between items observed by the same user, we achieve the superior user preference on item content for user-item interaction prediction. Observing the experimental results on three datasets, we demonstrate the effectiveness and efficiency of our LightGT.

Chapter 7
Research Frontiers

This book provides a comprehensive and systematic exploration of advancing recommender systems with graph convolutional networks (GCNs), with an in-depth focus on five critical challenges: the over-smoothing problem, the noisy neighboring nodes issue, interpretability limitations, the attribute missing problem and the feature misalignment problem. These challenges represent significant constraints not only on the practical effectiveness of GCNs in recommender systems but also on the broader ability of recommendation models to handle real-world complexities. To overcome these challenges, this book introduces several targeted and innovative approaches aimed at enhancing the application and performance of GCNs in recommender systems.

Specifically, to address the over-smoothing problem, this book explores the interest-aware message-passing GCN model. This model leverages high-order neighbors within carefully constructed subgraphs, where user nodes with similar interests are grouped. This design enables more effective and localized information propagation. Then, the ClusterGCF model is proposed to address the problem of aggregating unreliable information. This model performs high-order graph convolution within cluster-specific graphs, which are constructed by identifying shared interests across multiple dimensions. This approach ensures the extraction of reliable and relevant information from high-order neighbors. Next, to enhance the interpretability of GCN-based recommendation methods, the SAGCN framework combines the strengths of large language models (LLMs) with conventional recommendation techniques. By extracting semantic aspects from user reviews, SAGCN provides interpretable and robust recommendations. To solve the attribute missing problem, this book introduces the A^2-GCN, which incorporates an attention mechanism to estimate the importance of different attributes and effectively utilize attribute information to learn user and item representations. Finally, the LightGT model, a Transformer-based framework, is proposed to address feature misalignment. LightGT integrates modal-specific embeddings and a layer-wise position encoder to efficiently extract multimodal features. To support reproducibility and further research, all datasets and source codes related to these models have been made publicly available, with links provided in the respective chapters.

F. Liu, L. Nie, *Advancing Recommender Systems with Graph Convolutional Networks*,
https://doi.org/10.1007/978-3-031-85093-6_7

Despite the extensive discussion and proposed solutions for the key technical challenges in recommender systems with GCNs, we recognize that this research area is still at an up-and-coming stage, with considerable opportunities for further exploration. In this regard, the book outlines several promising directions and challenges, aiming to inspire and guide future research in this evolving domain.

7.1 Dynamic and Streaming Graphs

Traditional GCN-based recommendation methods generally rely on static graphs constructed from user-item interactions. However, in real-world scenarios, user-item interactions evolve continuously over time, with new users and items emerging, resulting in dynamic graphs. In dynamic recommendation scenarios, such as social media and e-commerce platforms, simultaneously accounting for the user-item interactions and the evolution of dynamic graphs is essential for dynamically modeling user preferences across different time periods. This is crucial to ensure that recommendations results remain in synchrony with real-time changes in user behavior.

Compared with traditional static graph-based methods, dynamic graphs pose several severe challenges for the application of GCNs: (1) Long-term information forgetting. The continuous updates of dynamic interactions can lead to the forgetting of critical early-stage information, making it difficult for the model to accurately capture user preferences over extended time spans. (2) Dynamic preference modeling. A key challenge is developing methods to efficiently and rapidly update the model as the graph structure evolves, ensuring the timely and dynamic capture of emerging trends in user preferences. (2) Training efficiency. In dynamic scenarios, enabling online updates of user and item representations without retraining the entire graph is a critical challenge. This remains a major barrier to the practical deployment of dynamic graphs in real-world applications.

7.2 Scalability to Large-Scale Graphs

Many commercial recommender systems often have millions or even billions of users and items, producing extremely large graphs. Existing GCN-based recommendation methods already show disadvantages in computational complexity and model size compared to other types of recommendation methods when applied to smaller-scale experimental data. Therefore, directly applying GCNs to large-scale graphs would further exacerbate the computational and memory demands of recommendation systems.

Specifically, at first, traditional GCN-based methods require processing all neighbors (and neighbors of neighbors) during message passing, leading to an exponential increase in computational complexity when applied to large-scale graphs. Furthermore, storing and updating high-dimensional embeddings for billions of nodes demands substantial memory resources, often exceeding the capacity of typical hardware. These limitations severely restrict the practical application of GCNs in commercial scenarios. Addressing these challenges by reducing the computational and memory requirements of GCNs for large-scale graph processing is a critical and valuable research direction.

7.3 Multi-behavior and Multi-intent Modeling

In traditional GCN-based recommendation methods, edges in the graph typically represent a single type of interaction (*e.g.*, purchases). However, in real-world scenarios, there are multiple types of interactions between users and items (*e.g.*, clicks, likes, shares and purchases). The multi-behavior interactions can form a multi-relational graph, where edges between users and items represent different types of behaviors. Multi-relational graphs can provide a more nuanced representation of the relationships between users and items and can also alleviate data sparsity issues by incorporating diverse interaction data. Effectively leveraging the information within these graphs can significantly enhance the representation learning of users and items.

However, accurately modeling user preferences on multi-relational graphs poses significant challenges, primarily due to behavioral heterogeneity and the complexity of intent modeling. First, traditional GCN-based methods treat all edges uniformly, but modeling multiple behaviors on multi-relational graphs requires distinguishing between different types of relationships based on their characteristics. For example, purchasing behavior typically reflects stronger user intent compared to clicking behavior. Identifying effective strategies to address these distinctions remains a critical area for future research. Additionally, users often exhibit multiple concurrent or evolving intents (*e.g.*, purchasing while sharing), making it difficult to model these coexisting or shifting behaviors effectively on multi-relational graphs.

References

1. Arora, S., Liang, Y., Ma, T.: A simple but tough-to-beat baseline for sentence embeddings. In: Proceedings of the International Conference on Learning Representations, pp. 1–16. OpenReview.net (2017)
2. Atwood, J., Towsley, D.: Diffusion-convolutional neural networks. In: Proceedings of the 30th International Conference on Neural Information Processing Systems, p. 2001–2009. Curran Associates Inc. (2016). DOI 10.5555/3157096.3157320
3. Bauman, K., Liu, B., Tuzhilin, A.: Aspect based recommendations: Recommending items with the most valuable aspects based on user reviews. In: SIGKDD, p. 717–725. ACM (2017)
4. Bell, R.M., Koren, Y.: Lessons from the netflix prize challenge. ACM SIGKDD Explorations Newsletter **9**(2), 75–79 (2007). DOI 10.1145/1345448.1345465
5. van den Berg, R., Kipf, T.N., Welling, M.: Graph convolutional matrix completion. In: ICSCA. ACM (2018)
6. Blei, D.M., Ng, A.Y., Jordan, M.I.: Latent dirichlet allocation. Journal of machine Learning research **3**(Jan), 993–1022 (2003)
7. Bruna, J., Zaremba, W., Szlam, A., LeCun, Y.: Spectral networks and locally connected networks on graphs. In: Proceedings of International Conference on Learning Representations, pp. 1 – 14 (2014). DOI https://doi.org/10.48550/arXiv.1312.6203
8. Catherine, R., Cohen, W.: Transnets: Learning to transform for recommendation. In: Proceedings of the ACM Conference on Recommender Systems, p. 288–296. ACM (2017). DOI 10.1145/3109859.3109878
9. Chen, C., Wang, C., Tsai, M., Yang, Y.: Collaborative similarity embedding for recommender systems. In: Proceedings of the Web Conference, pp. 2637–2643. ACM (2019). DOI 10.1145/3308558.3313493
10. Chen, C., Zhang, M., Liu, Y., Ma, S.: Neural attentional rating regression with review-level explanations. In: Proceedings of the Web Conference, pp. 1583–1592. IW3C2 (2018). DOI 10.1145/3178876.3186070
11. Chen, C., Zhang, M., Ma, W., Liu, Y., Ma, S.: Efficient non-sampling factorization machines for optimal context-aware recommendation. In: Proceedings of the Web Conference, p. 2400–2410. ACM (2020). DOI 10.1145/3366423.3380303
12. Chen, F., Wang, J., Wei, Y., Zheng, H.T., Shao, J.: Breaking isolation: Multimodal graph fusion for multimedia recommendation by edge-wise modulation. In: Proceedings of the 30th ACM International Conference on Multimedia, pp. 385–394. ACM (2022). DOI 10.1145/3503161.3548399
13. Chen, H., Chen, Y., Wang, X., Xie, R., Wang, R., Xia, F., Zhu, W.: Curriculum disentangled recommendation with noisy multi-feedback. In: NeurIPS. CAI. (2024)
14. Chen, J., Zhang, H., He, X., Nie, L., Liu, W., Chua, T.S.: Attentive collaborative filtering: Multimedia recommendation with item-and component-level attention. In: Proceedings of

F. Liu, L. Nie, *Advancing Recommender Systems with Graph Convolutional Networks*,
https://doi.org/10.1007/978-3-031-85093-6

the International ACM SIGIR Conference on Research and Development in Information Retrieval, pp. 335–344. ACM (2017). DOI 10.1145/3077136.3080797
15. Chen, J., Zhuang, F., Hong, X., Ao, X., Xie, X., He, Q.: Attention-driven factor model for explainable personalized recommendation. In: Proceedings of the International ACM SIGIR Conference on Research and Development in Information Retrieval, pp. 909–912. ACM (2018). DOI 10.1145/3209978.3210083
16. Chen, L., Wu, L., Hong, R., Zhang, K., Wang, M.: Revisiting graph based collaborative filtering: A linear residual graph convolutional network approach. In: The AAAI Conference on Artificial Intelligence, pp. 27–34. AAAI Press (2020)
17. Chen, Q., Zhao, H., Li, W., Huang, P., Ou, W.: Behavior sequence transformer for e-commerce recommendation in alibaba. In: Proceedings of the International Workshop on Deep Learning Practice for High-Dimensional Sparse Data. ACM (2019). DOI 10.1145/3326937.3341261
18. Chen, X., Chen, H., Xu, H., Zhang, Y., Cao, Y., Qin, Z., Zha, H.: Personalized fashion recommendation with visual explanations based on multimodal attention network: Towards visually explainable recommendation. In: Proceedings of the International ACM SIGIR Conference on Research and Development in Information Retrieval, pp. 765–774. ACM (2019). DOI 10.1145/3331184.3331254
19. Cheng, H.T., Koc, L., Harmsen, J., Shaked, T., Chandra, T., Aradhye, H., Anderson, G., Corrado, G., Chai, W., Ispir, M., et al.: Wide & deep learning for recommender systems. In: Proceedings of the Workshop on Deep Learning for Recommender Systems, pp. 7–10. ACM (2016). DOI 10.1145/2988450.2988454
20. Cheng, Z., Chang, X., Zhu, L., Kanjirathinkal, R.C., Kankanhalli, M.: Mmalfm: Explainable recommendation by leveraging reviews and images. ACM Transactions on Information Systems **37**(2), 16 (2019). DOI 10.1145/3291060
21. Cheng, Z., Ding, Y., He, X., Zhu, L., Song, X., Kankanhalli, M.: A^3ncf: An adaptive aspect attention model for rating prediction. In: Proceedings of the International Joint Conference on Artificial Intelligence, pp. 3748–3754. AAAI Press (2018)
22. Cheng, Z., Ding, Y., Zhu, L., Kankanhalli, M.: Aspect-aware latent factor model: Rating prediction with ratings and reviews. In: Proceedings of the Web Conference, pp. 639–648. IW3C2 (2018). DOI 10.1145/3178876.3186145
23. Cheng, Z., Dong, J., Liu, F., Zhu, L., Yang, X., Wang, M.: Disentangled cascaded graph convolution networks for multi-behavior recommendation. ACM Transactions on Recommender Systems **2**(4) (2024). DOI 10.1145/3673244
24. Cheng, Z., Han, S., Liu, F., Zhu, L., Gao, Z., Peng, Y.: Multi-behavior recommendation with cascading graph convolution networks. In: Proceedings of the Web Conference, p. 1181–1189. ACM (2023). DOI 10.1145/3543507.3583439
25. Cheng, Z., Liu, F., Mei, S., Guo, Y., Zhu, L., Nie, L.: Feature-level attentive icf for recommendation. ACM Transactions on Information Systems **40**(4), 1–24 (2022). DOI 10.1145/3490477
26. Chin, J.Y., Zhao, K., Joty, S., Cong, G.: ANR: Aspect-based neural recommender. In: Proceedings of the ACM International Conference on Information & Knowledge Management, pp. 147–156. ACM (2018). DOI 10.1145/3269206.3271810
27. Christoffel, F., Paudel, B., Newell, C., Bernstein, A.: Blockbusters and wallflowers: Accurate, diverse, and scalable recommendations with random walks. In: Proceedings of the ACM Conference on Recommender Systems, pp. 163–170. ACM (2015). DOI 10.1145/2792838.2800180
28. Da'u, A., Salim, N.: Recommendation system based on deep learning methods: a systematic review and new directions. Artificial Intelligence Review **53**(4), 2709–2748 (2022)
29. Defferrard, M., Bresson, X., Vandergheynst, P.: Convolutional neural networks on graphs with fast localized spectral filtering. In: Proceedings of the 30th International Conference on Neural Information Processing Systems, p. 3844–3852. Curran Associates Inc. (2016). DOI 10.5555/3157382.3157527
30. Deshpande, M., Karypis, G.: Item-based top-n recommendation algorithms. In: ACM Transactions on Information Systems, pp. 143–177. ACM (2004). DOI 10.1145/963770.963776
31. Devlin, J., Chang, M.W., Lee, K., Toutanova, K.: Bert: Pre-training of deep bidirectional transformers for language understanding. In: Proceedings of the Conference of the North

American Chapter of the Association for Computational Linguistics, pp. 4171–4186. ACL (2019). DOI 10.18653/V1/N19-1423
32. Ding, Y., Ma, Y., Wong, W.K., Chua, T.: Modeling instant user intent and content-level transition for sequential fashion recommendation. IEEE Transactions on Multimedia **24**, 2687–2700 (2022). DOI 10.1109/TMM.2021.3088281
33. Du, Y., Wei, Y., Ji, W., Liu, F., Luo, X., Nie, L.: Multi-queue momentum contrast for microvideo-product retrieval. In: Proceedings of the ACM International Conference on Web Search and Data Mining, p. 1003–1011. ACM (2023). DOI 10.1145/3539597.3570405
34. Duvenaud, D.K., Maclaurin, D., Iparraguirre, J., Bombarell, R., Hirzel, T., Aspuru-Guzik, A., Adams, R.P.: Convolutional networks on graphs for learning molecular fingerprints. In: C. Cortes, N. Lawrence, D. Lee, M. Sugiyama, R. Garnett (eds.) Advances in Neural Information Processing Systems, vol. 28, pp. 1–9. Curran Associates, Inc. (2015)
35. Ebesu, T., Shen, B., Fang, Y.: Collaborative memory network for recommendation systems. In: Proceedings of the International ACM SIGIR Conference on Research and Development in Information Retrieval, pp. 515–524. ACM (2018). DOI 10.1145/3209978.3209991
36. Fan, S., Zhu, J., Han, X., Shi, C., Hu, L., Ma, B., Li, Y.: Metapath-guided heterogeneous graph neural network for intent recommendation. In: Proceedings of the ACM SIGKDD International Conference on Knowledge Discovery & Data Mining, p. 2478–2486. ACM (2019). DOI 10.1145/3292500.3330673
37. Fan, W., Liu, X., Jin, W., Zhao, X., Tang, J., Li, Q.: Graph trend filtering networks for recommendation. In: Proceedings of the 45th International ACM SIGIR Conference on Research and Development in Information Retrieval, pp. 112–121 (2022)
38. Fan, W., Ma, Y., Li, Q., He, Y., Zhao, Y.E., Tang, J., Yin, D.: Graph neural networks for social recommendation. In: Proceedings of the Web Conference, pp. 417–426. ACM (2019). DOI 10.1145/3308558.3313488
39. Fan, Z., Liu, Z., Zhang, J., Xiong, Y., Zheng, L., Yu, P.S.: Continuous-time sequential recommendation with temporal graph collaborative transformer. In: Proceedings of the ACM International Conference on Information & Knowledge Management, pp. 433–442. ACM (2021). DOI 10.1145/3459637.3482242
40. Fei, H., Wu, S., Li, J., Li, B., Li, F., Qin, L., Zhang, M., Zhang, M., Chua, T.S.: Lasuie: Unifying information extraction with latent adaptive structure-aware generative language model. In: Proceedings of the Advances in Neural Information Processing Systems, pp. 15460–15475. CAI (2022). DOI abs/10.5555/3600270.3601395
41. Fei, H., Zhang, M., Ji, D.: Cross-lingual semantic role labeling with high-quality translated training corpus. In: Proceedings of the Annual Meeting of the Association for Computational Linguistics, pp. 7014–7026. ACL (2020). DOI 10.18653/v1/2020.acl-main.627
42. Fouss, F., Pirotte, A., Renders, J., Saerens, M.: Random-walk computation of similarities between nodes of a graph with application to collaborative recommendation. IEEE Transactions on Knowledge and Data Engineering **19**(3), 355–369 (2007). DOI 10.1109/TKDE.2007.46
43. Fouss, F., Pirotte, A., Renders, J., Saerens, M.: Random-walk computation of similarities between nodes of a graph with application to collaborative recommendation. IEEE Transactions on Knowledge and Data Engineering **19**(3), 355–369 (2007). DOI 10.1109/TKDE.2007.46
44. Gao, Y., Sheng, T., Xiang, Y., Xiong, Y., Wang, H., Zhan, J.: Chat-rec: Towards interactive and explainable llms-augmented recommender system. CoRR **abs/2305.07001** (2023)
45. Geng, S., Liu, S., Fu, Z., Ge, Y., Zhang, Y.: Recommendation as language processing (rlp): A unified pretrain, personalized prompt & predict paradigm (p5). In: Proceedings of the ACM Conference on Recommender Systems, p. 299–315. ACM (2022). DOI 10.1145/3523227.3546767
46. Gori, M., Pucci, A.: Itemrank: A random-walk based scoring algorithm for recommender engines. In: Proceedings of the International Joint Conference on Artificial Intelligence, pp. 2766–2771. AAAI Press (2007). DOI 10.5555/1625275.1625720
47. Guo, G., Zhang, J., Yorke-Smith, N.: A novel recommendation model regularized with user trust and item ratings. IEEE Transactions on Knowledge and Data Engineering **28**(7), 1607–1620 (2016). DOI 10.1109/TKDE.2016.2528249

48. Hamilton, W.L., Ying, Z., Leskovec, J.: Inductive representation learning on large graphs. In: Proceedings of the International Conference on Neural Information Processing Systems, pp. 1024–1034. CAI. (2017). DOI 10.5555/3294771.3294869
49. Han, K., Xiao, A., Wu, E., Guo, J., Xu, C., Wang, Y.: Transformer in transformer. Proceedings of the International Conference on Neural Information Processing Systems **34**, 15908–15919 (2021). DOI 10.5555/3540261.3541478
50. He, K., Zhang, X., Ren, S., Sun, J.: Deep residual learning for image recognition. In: Proceedings of the IEEE/CVF Conference on Computer Vision and Pattern Recognition, pp. 770–778. IEEE Computer Society (2016). DOI 10.1109/CVPR.2016.90
51. He, R., McAuley, J.: Vbpr: visual bayesian personalized ranking from implicit feedback. In: Proceedings of the AAAI Conference on Artificial Intelligence, pp. 144–150. AAAI Press (2016). DOI 10.5555/3015812.3015834
52. He, X., Chen, T., Kan, M.Y., Chen, X.: Trirank: Review aware explainable recommendation by modeling aspects. In: Proceedings of the ACM International on Conference on Information and Knowledge Management, pp. 1661–1670. ACM (2015). DOI 10.1145/2806416.2806504
53. He, X., Chua, T.: Neural factorization machines for sparse predictive analytics. In: Proceedings of the International ACM SIGIR Conference on Research and Development in Information Retrieval, p. 355–364. ACM (2017). DOI 10.1145/3077136.3080777
54. He, X., Deng, K., Wang, X., Li, Y., Zhang, Y., Wang, M.: Lightgcn: Simplifying and powering graph convolution network for recommendation. In: Proceedings of the International ACM SIGIR Conference on Research and Development in Information Retrieval, p. 639–648. ACM (2020). DOI 10.1145/3397271.3401063
55. He, X., He, Z., Song, J., Liu, Z., Jiang, Y.G., Chua, T.S.: NAIS: Neural attentive item similarity model for recommendation. IEEE Transactions on Knowledge and Data Engineering **30**, 2354–2366 (2018). DOI 10.1109/TKDE.2018.2831682
56. He, X., Liao, L., Zhang, H., Nie, L., Hu, X., Chua, T.S.: Neural collaborative filtering. In: Proceedings of the International Conference on World Wide Web, pp. 173–182. IW3C2 (2017). DOI 10.1145/3038912.3052569
57. Hershey, S., Chaudhuri, S., Ellis, D.P., Gemmeke, J.F., Jansen, A., Moore, R.C., Plakal, M., Platt, D., Saurous, R.A., Seybold, B., et al.: Cnn architectures for large-scale audio classification. In: Proceedings of the IEEE International Conference on Acoustics, Speech and Signal Processing, pp. 131–135. IEEE (2017). DOI 10.1109/ICASSP.2017.7952132
58. Hochreiter, S., Schmidhuber, J.: Long short-term memory. Neural Computation **9**(8), 1735–1780 (1997). DOI 10.1162/neco.1997.9.8.1735
59. Hsieh, C.K., Yang, L., Cui, Y., Lin, T.Y., Belongie, S., Estrin, D.: Collaborative metric learning. In: Proceedings of the International Conference on World Wide Web, pp. 193–201. IW3C2 (2017). DOI 10.1145/3038912.3052639
60. Hu, Y., Koren, Y., Volinsky, C.: Collaborative filtering for implicit feedback datasets. In: Proceedings of the International Conference on Data Mining, pp. 263–272. IEEE (2008). DOI 10.1109/ICDM.2008.22
61. Hu, Y., Yang, C., Zhan, P., Zhao, J., Li, Y., Li, X.: Efficient continuous knn join processing for real-time recommendation. In: Personal and Ubiquitous Computing, vol. 24, pp. 1–11 (2019). DOI 10.1007/S00779-019-01282-5
62. Hu, Y., Zhan, P., Xu, Y., Zhao, J., Li, Y., Li, X.: Temporal representation learning for time series classification. Neural Computing and Applications **32**, 1–14 (2020). DOI 10.1007/S00521-020-05179-W
63. Huang, Q., Lei, Y., Tung, A.K.: Point-to-hyperplane nearest neighbor search beyond the unit hypersphere. In: Proceedings of the 2021 International Conference on Management of Data, pp. 777–789 (2021)
64. Jang, E., Gu, S., Poole, B.: Categorical reparameterization with gumbel-softmax. In: Proceedings of the International Conference on Learning Representations, pp. 1–12. OpenReview.net (2017)
65. Ji, J., Li, Z., Xu, S., Hua, W., Ge, Y., Tan, J., Zhang, Y.: Genrec: Large language model for generative recommendation. CoRR **abs/2307.00457** (2023)

66. Ji, W., Chen, L., Wei, Y., Wu, Y., Chua, T.S.: Mrtnet: Multi-resolution temporal network for video sentence grounding. In: Proceedings of the IEEE International Conference on Acoustics, Speech and Signal Processing, pp. 2770–2774 (2024). DOI 10.1109/ICASSP48485.2024.10447846
67. Ji, W., Liang, R., Zheng, Z., Zhang, W., Zhang, S., Li, J., Li, M., Chua, T.s.: Are binary annotations sufficient? video moment retrieval via hierarchical uncertainty-based active learning. In: Proceedings of the IEEE/CVF Conference on Computer Vision and Pattern Recognition, pp. 23013–23022. IEEE (2023). DOI 10.1109/CVPR52729.2023.02204
68. Ji, W., Liu, X., Zhang, A., Wei, Y., Ni, Y., Wang, X.: Online distillation-enhanced multi-modal transformer for sequential recommendation. In: Proceedings of the ACM International Conference on Multimedia, pp. 955–965. ACM (2023). DOI 10.1145/3581783.3612091
69. Kang, W., Ni, J., Mehta, N., Sathiamoorthy, M., Hong, L., Chi, E.H., Cheng, D.Z.: Do llms understand user preferences? evaluating llms on user rating prediction. CoRR **abs/2305.06474** (2023)
70. Kingma, D.P., Ba, J.: Adam: A method for stochastic optimization. In: Proceedings of the International Conference on Learning Representations, pp. 1–15. OpenReview.net (2015)
71. Kipf, T.N., Welling, M.: Semi-supervised classification with graph convolutional networks. In: Proceedings of the International Conference on Learning Representations, pp. 1–14. OpenReview.net (2017)
72. Kipf, T.N., Welling, M.: Semi-supervised classification with graph convolutional networks. In: International Conference on Learning Representations (2017). URL `https://openreview.net/forum?id=SJU4ayYgl`
73. Klaus, S., Van Hecke, R., Djafari Naini, K., Altingovde, I.S., Bernabé-Moreno, J., Herrera-Viedma, E.: Summarizing legal regulatory documents using transformers. In: Proceedings of the International ACM SIGIR Conference on Research and Development in Information Retrieval, pp. 2426–2430. ACM (2022). DOI 10.1145/3477495.3531872
74. Kolesnikov, A., Dosovitskiy, A., Weissenborn, D., Heigold, G., Uszkoreit, J., Beyer, L., Minderer, M., Dehghani, M., Houlsby, N., Gelly, S., Unterthiner, T., Zhai, X.: An image is worth 16x16 words: Transformers for image recognition at scale. In: Proceedings of the International Conference on Learning Representations, pp. 1–21. OpenReview.net (2021)
75. Koren, Y.: Factorization meets the neighborhood: a multifaceted collaborative filtering model. In: Proceedings of the ACM SIGKDD International Conference on Knowledge Discovery & Data Mining, p. 426–434. ACM (2008). DOI 10.1145/1401890.1401944
76. Koren, Y., Bell, R., Volinsky, C.: Matrix factorization techniques for recommender systems. Computer **42**(8), 30–37 (2009). DOI 10.1109/MC.2009.263
77. Li, A., Cheng, Z., Liu, F., Gao, Z., Guan, W., Peng, Y.: Disentangled graph neural networks for session-based recommendation. IEEE Transactions on Knowledge and Data Engineering **35**(8), 7870–7882 (2023). DOI 10.1109/TKDE.2022.3208782
78. Li, C., Quan, C., Peng, L., Qi, Y., Deng, Y., Wu, L.: A capsule network for recommendation and explaining what you like and dislike. In: Proceedings of the International ACM SIGIR Conference on Research and Development in Information Retrieval, p. 275–284. ACM (2019). DOI 10.1145/3331184.3331216
79. Li, L., Zhang, Y., Chen, L.: Personalized prompt learning for explainable recommendation. ACM Transactions on Information Systems **41**(4), 103:1–103:26 (2023). DOI 10.1145/3580488
80. Li, Y., Zhu, J., Liu, W., Su, L., Cai, G., Zhang, Q., Tang, R., Xiao, X., He, X.: Pear: Personalized re-ranking with contextualized transformer for recommendation. In: Proceedings of the International Conference on World Wide Web, p. 62–66. ACM (2022). DOI 10.1145/3487553.3524208
81. Li, Z., Liu, F., Wei, Y., Cheng, Z., Nie, L., Kankanhalli, M.: Attribute-driven disentangled representation learning for multimodal recommendation. In: Proceedings of the ACM International Conference on Multimedia, p. 9660–9669. ACM (2024). DOI 10.1145/3664647.3681148

82. Lin, Z., Tian, C., Hou, Y., Zhao, W.X.: Improving graph collaborative filtering with neighborhood-enriched contrastive learning. In: Proceedings of the ACM Web Conference 2022, pp. 2320–2329 (2022)
83. Liu, F., Chen, H., Cheng, Z., Liu, A., Nie, L., Kankanhalli, M.: Disentangled multimodal representation learning for recommendation. IEEE Transactions on Multimedia **25**, 7149–7159 (2023). DOI 10.1109/TMM.2022.3217449
84. Liu, F., Chen, H., Cheng, Z., Nie, L., Kankanhalli, M.: Semantic-guided feature distillation for multimodal recommendation. In: Proceedings of the ACM International Conference on Multimedia, p. 6567–6575. ACM (2023). DOI 10.1145/3581783.3611886
85. Liu, F., Cheng, Z., Chen, H., Wei, Y., Nie, L., Kankanhalli, M.: Privacy-preserving synthetic data generation for recommendation systems. In: Proceedings of the International ACM SIGIR Conference on Research and Development in Information Retrieval, p. 1379–1389. ACM (2022). DOI 10.1145/3477495.3532044
86. Liu, F., Cheng, Z., Sun, C., Wang, Y., Nie, L., Kankanhalli, M.: User diverse preference modeling by multimodal attentive metric learning. In: Proceedings of the ACM International Conference on Multimedia, p. 1526–1534. ACM (2019). DOI 10.1145/3343031.3350953
87. Liu, F., Cheng, Z., Zhu, L., Gao, Z., Nie, L.: Interest-aware message-passing gcn for recommendation. In: Proceedings of the Web Conference, p. 1296–1305. ACM (2021). DOI 10.1145/3442381.3449986
88. Liu, F., Cheng, Z., Zhu, L., Liu, C., Nie, L.: An attribute-aware attentive gcn model for attribute missing in recommendation. IEEE Transactions on Knowledge and Data Engineering **34**(9), 4077–4088 (2022). DOI 10.1109/TKDE.2020.3040772
89. Liu, F., Liu, Y., Chen, H., Cheng, Z., Nie, L., Kankanhalli, M.: Understanding before recommendation: Semantic aspect-aware review exploitation via large language models. ACM Transactions on Information Systems pp. 1–26 (2024). DOI 10.1145/3704999
90. Liu, F., Zhao, S., Cheng, Z., Nie, L., Kankanhalli, M.: Cluster-based graph collaborative filtering. ACM Transactions on Information Systems **42**(6), 1–24 (2024). DOI 10.1145/3687481
91. Liu, H., Wei, Y., Liu, F., Wang, W., Nie, L., Chua, T.S.: Multimodal fusion via meta-learning towards micro-video recommendation. ACM Transactions on Information Systems **42**(2), 1–26 (2023). DOI 10.1145/3617827
92. Liu, H., Wei, Y., Yin, J., Nie, L.: Hs-gcn: Hamming spatial graph convolutional networks for recommendation. IEEE Transactions on Knowledge and Data Engineering **35**(6), 5977–5990 (2023). DOI 10.1109/TKDE.2022.3158317
93. Liu, M., Zhang, F., Luo, X., Liu, F., Wei, Y., Nie, L.: Advancing video question answering with a multi-modal and multi-layer question enhancement network. In: Proceedings of the ACM International Conference on Multimedia, p. 3985–3993. ACM (2023). DOI 10.1145/3581783.3612239
94. Liu, Q., Chen, N., Sakai, T., Wu, X.: A first look at llm-powered generative news recommendation. CoRR **abs/2305.06566** (2023)
95. Liu, S., Chen, Z., Liu, H., Hu, X.: User-video co-attention network for personalized micro-video recommendation. In: The World Wide Web Conference, pp. 3020–3026 (2019)
96. Liu, X., Wu, W.: Learning context-aware latent representations for context-aware collaborative filtering. In: Proceedings of the 38th International ACM SIGIR Conference on Research and Development in Information Retrieval, p. 887–890. Association for Computing Machinery (2015)
97. Liu, Y., Ren, Z., Zhang, W.N., Che, W., Liu, T., Yin, D.: Keywords generation improves e-commerce session-based recommendation. In: Proceedings of The Web Conference 2020, p. 1604–1614. ACM (2020)
98. Liu, Y., Yang, S., Lei, C., Wang, G., Tang, H., Zhang, J., Sun, A., Miao, C.: Pre-training graph transformer with multimodal side information for recommendation. In: Proceedings of the ACM International Conference on Multimedia, pp. 2853–2861. ACM (2021). DOI 10.1145/3474085.3475709

99. Ma, J., Zhou, C., Cui, P., Yang, H., Zhu, W.: Learning disentangled representations for recommendation. In: Proceedings of the International Conference on Neural Information Processing Systems, pp. 5712–5723. CAI. (2019)
100. Maas, A.L., Hannun, A.Y., Ng, A.Y.: Rectifier nonlinearities improve neural network acoustic models. In: Proceedings of the International Conference on Machine Learning (2013)
101. van der Maaten, L., Hinton, G.: Visualizing data using t-SNE. Journal of Machine Learning Research **9**, 2579–2605 (2008)
102. Mao, K., Zhu, J., Xiao, X., Lu, B., Wang, Z., He, X.: Ultragcn: ultra simplification of graph convolutional networks for recommendation. In: Proceedings of the 30th ACM International Conference on Information & Knowledge Management, pp. 1253–1262. ACM (2021)
103. McAuley, J., Leskovec, J.: Hidden factors and hidden topics: understanding rating dimensions with review text. In: Proceedings of the ACM Conference on Recommender Systems, pp. 165–172. ACM (2013). DOI 10.1145/2507157.2507163
104. McAuley, J., Targett, C., Shi, Q., Van Den Hengel, A.: Image-based recommendations on styles and substitutes. In: Proceedings of the International ACM SIGIR Conference on Research and Development in Information Retrieval, pp. 43–52. ACM (2015). DOI 10.1145/2766462.2767755
105. Mikolov, T., Chen, K., Corrado, G., Dean, J.: Efficient estimation of word representations in vector space. In: Proceedings of the International Conference on Learning Representations. OpenReview.net (2013)
106. Nembrini, R., Carugno, C., Ferrari Dacrema, M., Cremonesi, P.: Towards recommender systems with community detection and quantum computing. In: Proceedings of the ACM Conference on Recommender Systems, p. 579–585. ACM (2022). DOI 10.1145/3523227.3551478
107. Pasricha, R., McAuley, J.: Translation-based factorization machines for sequential recommendation. In: Proceedings of the ACM Conference on Recommender Systems, p. 63–71. ACM (2018). DOI 10.1145/3240323.3240356
108. Qin, Y., Fang, Y., Luo, H., Zhao, F., Wang, C.: Next point-of-interest recommendation with auto-correlation enhanced multi-modal transformer network. In: Proceedings of the 45th International ACM SIGIR Conference on Research and Development in Information Retrieval, pp. 2612–2616. ACM (2022). DOI 10.1145/3477495.3531905
109. Qiu, J., Tang, J., Ma, H., Dong, Y., Wang, K., Tang, J.: Deepinf: Social influence prediction with deep learning. In: Proceedings of the ACM SIGKDD International Conference on Knowledge Discovery & Data Mining, pp. 2110–2119. ACM (2018). DOI 10.1145/3219819.3220077
110. Ren, X., Wei, W., Xia, L., Su, L., Cheng, S., Wang, J., Yin, D., Huang, C.: Representation learning with large language models for recommendation. In: Proceedings of the ACM on Web Conference 2024, pp. 3464–3475 (2024)
111. Rendle, S.: Factorization machines. In: Proceedings of the International Conference on Data Mining, pp. 995–1000. IEEE (2011)
112. Rendle, S., Freudenthaler, C., Gantner, Z., Schmidt-Thieme, L.: BPR: Bayesian personalized ranking from implicit feedback. In: Proceedings of the Conference on Uncertainty in Artificial Intelligence, pp. 452–461. AUAI Press (2009)
113. Salakhutdinov, R., Mnih, A.: Probabilistic matrix factorization. In: Proceedings of the International Conference on Neural Information Processing Systems, p. 1257–1264. CAI. (2007). DOI 10.5555/2981562.2981720
114. Sandryhaila, A., Moura, J.M.: Big data analysis with signal processing on graphs: Representation and processing of massive data sets with irregular structure. IEEE Signal Processing Magazine **31**(5), 80–90 (2014). DOI 10.1109/MSP.2014.2329213
115. Sandryhaila, A., Moura, J.M.F.: Discrete signal processing on graphs. IEEE Transactions on Signal Processing **61**(7), 1644–1656 (2013). DOI 10.1109/TSP.2013.2238935
116. Shan, Y., Hoens, T.R., Jiao, J., Wang, H., Yu, D., Mao, J.: Deep crossing: Web-scale modeling without manually crafted combinatorial features. In: Proceedings of the ACM SIGKDD International Conference on Knowledge Discovery & Data Mining, p. 255–262. ACM (2016). DOI 10.1145/2939672.2939704

117. Shi, S., Zhang, M., Liu, Y., Ma, S.: Attention-based adaptive model to unify warm and cold starts recommendation. In: Proceedings of the ACM International Conference on Information & Knowledge Management, p. 127–136. ACM (2018). DOI 10.1145/3269206.3271710
118. Shuai, J., Zhang, K., Wu, L., Sun, P., Hong, R., Wang, M., Li, Y.: A review-aware graph contrastive learning framework for recommendation. In: SIGIR, p. 1283–1293. ACM (2022)
119. Shuman, D.I., Narang, S.K., Frossard, P., Ortega, A., Vandergheynst, P.: The emerging field of signal processing on graphs: Extending high-dimensional data analysis to networks and other irregular domains. IEEE Signal Processing Magazine **30**(3), 83–98 (2013). DOI 10.1109/MSP.2012.2235192
120. Sileo, D., Vossen, W., Raymaekers, R.: Zero-shot recommendation as language modeling. In: Advances in Information Retrieval: 44th European Conference on IR Research, ECIR 2022, Stavanger, Norway, April 10–14, 2022, Proceedings, Part II, p. 223–230. Springer (2022)
121. Simonyan, K., Zisserman, A.: Very deep convolutional networks for large-scale image recognition. In: Proceedings of the International Conference on Learning Representations, pp. 1–14. OpenReview.net (2015)
122. Sun, F., Liu, J., Wu, J., Pei, C., Lin, X., Ou, W., Jiang, P.: Bert4rec: Sequential recommendation with bidirectional encoder representations from transformer. In: Proceedings of the ACM international conference on information and knowledge management, pp. 1441–1450. ACM (2019). DOI 10.1145/3357384.3357895
123. Sun, J., Zhang, Y., Ma, C., Coates, M., Guo, H., Tang, R., He, X.: Multi-graph convolution collaborative filtering. In: Proceedings of IEEE International Conference on Data Mining, pp. 1306 – 1311 (2019)
124. Tan, Y., Zhang, M., Liu, Y., Ma, S.: Rating-boosted latent topics: Understanding users and items with ratings and reviews. In: Proceedings of the International Joint Conference on Artificial Intelligence, p. 2640–2646. AAAI Press (2016)
125. Vaswani, A., Shazeer, N., Parmar, N., Uszkoreit, J., Jones, L., Gomez, A.N., Kaiser, L., Polosukhin, I.: Attention is all you need. Advances in neural information processing systems **30** (2017)
126. Velickovic, P., Cucurull, G., Casanova, A., Romero, A., Liò, P., Bengio, Y.: Graph attention networks. In: Proceedings of the International Conference on Learning Representations, pp. 1–12. OpenReview.net (2019)
127. Wang, H., Leskovec, J.: Combining graph convolutional neural networks and label propagation. ACM Transactions on Information Systems **40**, 1–27 (2021). DOI 10.1145/3490478
128. Wang, H., Zhang, F., Wang, J., Zhao, M., Li, W., Xie, X., Guo, M.: Ripplenet: Propagating user preferences on the knowledge graph for recommender systems. In: Proceedings of the ACM International Conference on Information & Knowledge Management, pp. 417–426. ACM (2018). DOI 10.1145/3269206.3271739
129. Wang, H., Zhang, F., Wang, J., Zhao, M., Li, W., Xie, X., Guo, M.: Exploring high-order user preference on the knowledge graph for recommender systems. ACM Transactions on Information Systems **37**(3), 32:1–32:26 (2019). DOI 10.1145/3312738
130. Wang, S., Gan, T., Liu, Y., Wu, J., Cheng, Y., Nie, L.: Micro-influencer recommendation by multi-perspective account representation learning. IEEE Transactions on Multimedia **25**, 2749–2760 (2023). DOI 10.1109/TMM.2022.3151029
131. Wang, S., Wang, Y., Tang, J., Shu, K., Ranganath, S., Liu, H.: What your images reveal: Exploiting visual contents for point-of-interest recommendation. In: Proceedings of the Web Conference, pp. 391–400. IW3C2 (2017). DOI 10.1145/3038912.3052638
132. Wang, W., Feng, F., He, X., Zhang, H., Chua, T.S.: Clicks can be cheating: Counterfactual recommendation for mitigating clickbait issue. In: Proceedings of the International ACM SIGIR Conference on Research and Development in Information Retrieval, pp. 1288–1297 (2021). DOI 10.1145/3404835.3462962
133. Wang, W., Lin, X., Feng, F., He, X., Chua, T.S.: Generative recommendation: Towards next-generation recommender paradigm. arXiv pp. 1–25 (2023). DOI 10.48550/arXiv.2304.03516
134. Wang, W., Lin, X., Feng, F., He, X., Lin, M., Chua, T.S.: Causal representation learning for out-of-distribution recommendation. In: Proceedings of the Web Conference, pp. 3562–3571. ACM (2022). DOI 10.1145/3485447.3512251

135. Wang, X., He, X., Cao, Y., Liu, M., Chua, T.S.: KGAT: knowledge graph attention network for recommendation. In: Proceedings of the ACM SIGKDD International Conference on Knowledge Discovery & Data Mining, pp. 950–958. ACM (2019). DOI 10.1145/3292500.3330989
136. Wang, X., He, X., Wang, M., Feng, F., Chua, T.S.: Neural graph collaborative filtering. In: Proceedings of the International ACM SIGIR Conference on Research and Development in Information Retrieval, pp. 165–174. ACM (2019). DOI 10.1145/3331184.3331267
137. Wang, X., Ji, H., Shi, C., Wang, B., Ye, Y., Cui, P., Yu, P.S.: Heterogeneous graph attention network. In: Proceedings of the Web Conference, p. 2022–2032. ACM (2019). DOI 10.1145/3308558.3313562
138. Wang, X., Jin, H., Zhang, A., He, X., Xu, T., Chua, T.S.: Disentangled graph collaborative filtering. In: Proceedings of the International ACM SIGIR Conference on Research and Development in Information Retrieval, p. 1001–1010. ACM (2020). DOI 10.1145/3397271.3401137
139. Wang, Y., Tang, S., Lei, Y., Song, W., Wang, S., Zhang, M.: Disenhan: Disentangled heterogeneous graph attention network for recommendation. In: Proceedings of the ACM International Conference on Information & Knowledge Management, p. 1605–1614. ACM (2020). DOI 10.1145/3340531.3411996
140. Wei, Y., Cheng, Z., Yu, X., Zhao, Z., Zhu, L., Nie, L.: Personalized hashtag recommendation for micro-videos. In: Proceedings of the ACM International Conference on Multimedia, pp. 1446–1454. ACM (2019). DOI 10.1145/3343031.3350858
141. Wei, Y., Liu, W., Liu, F., Wang, X., Nie, L., Chua, T.S.: Lightgt: A light graph transformer for multimedia recommendation. In: Proceedings of the International ACM SIGIR Conference on Research and Development in Information Retrieval, p. 1508–1517. ACM (2023). DOI 10.1145/3539618.3591716
142. Wei, Y., Wang, X., Li, Q., Nie, L., Li, Y., Li, X., Chua, T.S.: Contrastive learning for cold-start recommendation. In: Proceedings of the 29th ACM International Conference on Multimedia, pp. 5382–5390. ACM (2021). DOI 10.1145/3474085.3475665
143. Wei, Y., Wang, X., Nie, L., He, X., Chua, T.S.: Graph-refined convolutional network for multimedia recommendation with implicit feedback. In: Proceedings of the ACM International Conference on Multimedia, pp. 3541–3549. ACM (2020). DOI 10.1145/3394171.3413556
144. Wei, Y., Wang, X., Nie, L., He, X., Hong, R., Chua, T.: MMGCN: multi-modal graph convolution network for personalized recommendation of micro-video. In: Proceedings of the ACM International Conference on Multimedia, pp. 1437–1445. ACM (2019). DOI 10.1145/3343031.3351034
145. Wu, F., Jr., A.H.S., Zhang, T., Fifty, C., Yu, T., Weinberger, K.Q.: Simplifying graph convolutional networks. In: Proceedings of the International Conference on Machine Learning, *Proceedings of Machine Learning Research*, vol. 97, pp. 6861–6871. PMLR (2019)
146. Wu, L., Li, S., Hsieh, C.J., Sharpnack, J.: Sse-pt: Sequential recommendation via personalized transformer. In: Proceedings of the ACM Conference on Recommender Systems, pp. 328–337. ACM (2020). DOI 10.1145/3383313.3412258
147. Wu, L., Zheng, Z., Qiu, Z., Wang, H., Gu, H., Shen, T., Qin, C., Zhu, C., Zhu, H., Liu, Q., Xiong, H., Chen, E.: A survey on large language models for recommendation. World Wide Web **27**, 1–34 (2023). DOI 10.1007/s11280-024-01291-2
148. Wu, Y., DuBois, C., Zheng, A.X., Ester, M.: Collaborative denoising auto-encoders for top-n recommender systems. In: Proceedings of the ACM International Conference on Web Search and Data Mining, pp. 153–162. ACM (2016). DOI 10.1145/2835776.2835837
149. Xavier, G., Yoshua, B.: Understanding the difficulty of training deep feedforward neural networks. In: Proceedings of the International Conference on Artificial Intelligence and Statistics, pp. 249–256. PMLR (2010)
150. Xia, L., Huang, C., Xu, Y., Dai, P., Zhang, B., Bo, L.: Multiplex behavioral relation learning for recommendation via memory augmented transformer network. In: Proceedings of the International ACM SIGIR Conference on Research and Development in Information Retrieval, pp. 2397–2406. ACM (2020). DOI 10.1145/3397271.3401445

151. Xia, L., Huang, C., Xu, Y., Dai, P., Zhang, X., Yang, H., Pei, J., Bo, L.: Knowledge-enhanced hierarchical graph transformer network for multi-behavior recommendation. In: Proceedings of the AAAI Conference on Artificial Intelligence, pp. 4486–4493 (2021). DOI 10.1609/aaai.v35i5.16576
152. Xiao, J., Ye, H., He, X., Zhang, H., Wu, F., Chua, T.S.: Attentional factorization machines: Learning the weight of feature interactions via attention networks. In: Proceedings of the International Joint Conference on Artificial Intelligence, p. 3119–3125. AAAI Press (2017). DOI 10.24963/IJCAI.2017/435
153. Xin, X., Karatzoglou, A., Arapakis, I., Jose, J.: Graph highway networks. ArXiv pp. 1–9 (2020). DOI 10.48550/arXiv.2004.04635
154. Xue, H., Dai, X., Zhang, J., Huang, S., Chen, J.: Deep matrix factorization models for recommender systems. In: Proceedings of the International Joint Conference on Artificial Intelligence, pp. 3203–3209. AAAI Press (2017). DOI 10.5555/3172077.3172336
155. Yan, M., Liu, F., Sun, J., Sun, F., Cheng, Z., Han, Y.: Behavior-contextualized item preference modeling for multi-behavior recommendation. In: Proceedings of the International ACM SIGIR Conference on Research and Development in Information Retrieval, p. 946–955. ACM (2024). DOI 10.1145/3626772.3657696
156. Yang, J., Chen, C., Wang, C., Tsai, M.: HOP-rec: high-order proximity for implicit recommendation. In: Proceedings of the ACM Conference on Recommender Systems, pp. 140–144. ACM (2018). DOI 10.1145/3240323.3240381
157. Ying, R., He, R., Chen, K., Eksombatchai, P., Hamilton, W.L., Leskovec, J.: Graph convolutional neural networks for web-scale recommender systems. In: Proceedings of the ACM SIGKDD International Conference on Knowledge Discovery & Data Mining, pp. 974–983. ACM (2018). DOI 10.1145/3219819.3219890
158. Yu, L., Zhang, C., Pei, S., Sun, G., Zhang, X.: Walkranker: A unified pairwise ranking model with multiple relations for item recommendation. In: Proceedings of the AAAI Conference on Artificial Intelligence, pp. 2596–2603. AAAI Press (2018). DOI abs/10.5555/3504035.3504352
159. Yuan, E., Guo, W., He, Z., Guo, H., Liu, C., Tang, R.: Multi-behavior sequential transformer recommender. In: Proceedings of the 45th International ACM SIGIR Conference on Research and Development in Information Retrieval, pp. 1642–1652. ACM (2022). DOI 10.1145/3477495.3532023
160. Yuan, L., Chen, Y., Wang, T., Yu, W., Shi, Y., Jiang, Z., Tay, F.E.H., Feng, J., Yan, S.: Tokens-to-token vit: Training vision transformers from scratch on imagenet. In: Proceedings of the International Conference on Computer Vision, pp. 538–547. IEEE (2021). DOI 10.1109/ICCV48922.2021.00060
161. Yue, Z., Rabhi, S., Moreira, G.d.S.P., Wang, D., Oldridge, E.: Llamarec: Two-stage recommendation using large language models for ranking. arXiv preprint arXiv:2311.02089 (2023)
162. Zhang, F., Yuan, N.J., Lian, D., Xie, X., Ma, W.Y.: Collaborative knowledge base embedding for recommender systems. In: Proceedings of the ACM SIGKDD International Conference on Knowledge Discovery & Data Mining, pp. 353–362. ACM (2016). DOI 10.1145/2939672.2939673
163. Zhang, H., McAuley, J.: Stacked mixed-order graph convolutional networks for collaborative filtering. In: Proceedings of the 2020 SIAM International Conference on Data Mining, pp. 73–81. Society for Industrial and Applied Mathematics (2020)
164. Zhang, J., Bao, K., Zhang, Y., Wang, W., Feng, F., He, X.: Is chatgpt fair for recommendation? evaluating fairness in large language model recommendation. In: Proceedings of the ACM Conference on Recommender Systems, p. 993–999. ACM (2023). DOI 10.1145/3604915.3608860
165. Zhang, J., Chow, C., Xu, J.: Enabling kernel-based attribute-aware matrix factorization for rating prediction. IEEE Transactions on Knowledge and Data Engineering **29**(4), 798–812 (2017). DOI 10.1109/TKDE.2016.2641439

166. Zhang, J., Xie, R., Hou, Y., Zhao, W.X., Lin, L., Wen, J.: Recommendation as instruction following: A large language model empowered recommendation approach. ACM Transactions on Information Systems pp. 1–36 (2023). DOI 10.1145/3708882
167. Zhang, J., Zhu, Y., Liu, Q., Wu, S., Wang, S., Wang, L.: Mining latent structures for multimedia recommendation. In: Proceedings of the ACM International Conference on Multimedia, p. 3872–3880. ACM (2021). DOI 10.1145/3474085.3475259
168. Zhang, Y., Ai, Q., Chen, X., Croft, W.B.: Joint representation learning for top-n recommendation with heterogeneous information sources. In: Proceedings of the ACM International Conference on Information & Knowledge Management, pp. 1449–1458. ACM (2017). DOI 10.1145/3132847.3132892
169. Zhang, Y., Cheng, Z., Liu, F., Yang, X., Peng, Y.: Decoupled domain-specific and domain-conditional representation learning for cross-domain recommendation. Information Processing & Management **61**(3), 103689 (2024). DOI 10.1016/j.ipm.2024.103689
170. Zhao, C., Li, C., Fu, C.: Cross-domain recommendation via preference propagation graphnet. In: Proceedings of the ACM International Conference on Information & Knowledge Management, p. 2165–2168. ACM (2019). DOI 10.1145/3357384.3358166
171. Zheng, L., Lu, C., Jiang, F., Zhang, J., Yu, P.S.: Spectral collaborative filtering. In: Proceedings of the ACM Conference on Recommender Systems, pp. 311–319. ACM (2018). DOI 10.1145/3240323.3240343
172. Zheng, L., Lu, C.T., He, L., Xie, S., Noroozi, V., Huang, H., Yu, P.S.: MARS: Memory attention-aware recommender system. In: Proceedings of the IEEE International Conference on Data Science and Advanced Analytics, pp. 11–20. IEEE (2019). DOI 10.1109/DSAA.2019.00015
173. Zheng, L., Noroozi, V., Yu, P.S.: Joint deep modeling of users and items using reviews for recommendation. In: Proceedings of the ACM International Conference on Web Search and Data Mining, pp. 425–434. ACM (2017). DOI 10.1145/3018661.3018665
174. Zhou, G., Zhu, X., Song, C., Fan, Y., Zhu, H., Ma, X., Yan, Y., Jin, J., Li, H., Gai, K.: Deep interest network for click-through rate prediction. In: Proceedings of the ACM SIGKDD International Conference on Knowledge Discovery & Data Mining, pp. 1059–1068. ACM (2018). DOI 10.1145/3219819.3219823
175. Zhou, X., Zhou, H., Liu, Y., Zeng, Z., Miao, C., Wang, P., You, Y., Jiang, F.: Bootstrap latent representations for multi-modal recommendation. In: Proceedings of the Web Conference, p. 845–854. ACM (2023). DOI 10.1145/3543507.3583251
176. Zou, J., Kanoulas, E., Ren, P., Ren, Z., Sun, A., Long, C.: Improving conversational recommender systems via transformer-based sequential modelling. In: Proceedings of the 45th International ACM SIGIR Conference on Research and Development in Information Retrieval, pp. 2319–2324. ACM (2022). DOI 10.1145/3477495.3531852

Made in the USA
Monee, IL
14 September 2026

62976165R00098